Beyond Prime Time

Daytime soap operas. Evening news. Late-night talk shows. Television has long been defined by its daily schedule and the viewing habits that develop around it. Technologies like DVRs, iPods, and online video have freed audiences from rigid time constraints—we no longer have to wait for a program to be "on" to watch it—but scheduling still plays a major role in the production of television.

Prime-time series programming between 8:00 and 11:00 p.m. has dominated most critical discussion about television since its beginnings, but *Beyond Prime Time* brings together leading television scholars to explore how shifts in television's industrial practices and new media convergence have affected the other 80 percent of the viewing day. The contributors explore a broad range of non-prime-time forms including talk shows, soap operas, news, syndication, and children's programs, non-series forms such as sports and made-for-television movies, as well as entities such as local affiliate stations and public television.

Importantly, all of these forms rely on norms of production, financing, and viewer habits that distinguish them from the practices common among prime-time series and often from each other. Each of the chapters examines how the production practices and textual strategies of a particular programming form have shifted in response to sweeping industry changes, together telling the story of a medium in transition at the beginning of the twenty-first century.

Contributors: Sarah Banet-Weiser, Victoria E. Johnson, Jeffrey P. Jones, Derek Kompare, Elana Levine, Amanda D. Lotz, Jonathan Nichols-Pethick, Laurie Ouellette, Erin Copple Smith.

Amanda D. Lotz is Associate Professor of Communication Studies at the University of Michigan. She is author of *Redesigning Women: Television after the Network Era* and *The Television will be Revolutionized*.

Beyond Prime Time

Television Programming in the Post-Network Era

Edited by
Amanda D. Lotz

NEW YORK AND LONDON

First published 2009
by Routledge
270 Madison Ave, New York, NY 10016

Simultaneously published in the UK
by Routledge
2 Park Square, Milton Park, Abingdon, Oxon OX14 4RN

Routledge is an imprint of the Taylor & Francis Group, an informa business

Typeset in Gill Sans and Perpetua by
Book Now Ltd, London
Printed and bound in the United States of America on acid-free paper by Walsworth Publishing Company, Marceline, MO

Library of Congress Cataloging in Publication Data
Beyond prime time: television programming in the post-network era / edited by Amanda Lotz.
p. cm.
Includes bibliographical references and index.
1. Television broadcasting—United States. I. Lotz, Amanda D., 1974–
PN1992.3.U5B45 2009
384.550973'0905—dc22 2009012258

ISBN10: 0–415–99668–6 (hbk)
ISBN10: 0–415–99669–4 (pbk)
ISBN10: 0–203–88450–7 (ebk)

ISBN13: 978–0–415–99668–6 (hbk)
ISBN13: 978–0–415–99669–3 (pbk)
ISBN13: 978–0–203–88450–8 (ebk)

With sincere gratitude to my contributors
ADL

Contents

Figures

Contributors

Sarah Banet-Weiser is Associate Professor in the Annenberg School for Communication at the University of Southern California. She is the author of *The Most Beautiful Girl in the World: Beauty Pageants and National Identity*, and *Kids Rule!: Nickelodeon and Consumer Citizenship*. She has co-edited, along with Cynthia Chris and Anthony Freitas, *Cable Visions: Television Beyond Broadcasting*.

Victoria E. Johnson is Associate Professor of Film and Media Studies and African American Studies at the University of California, Irvine. Her publications include *Heartland TV: Prime Time Television and the Struggle for U.S. Identity*, and chapters and articles in collections and journals such as *The Television Studies Reader*, *The Revolution Wasn't Televised: Sixties Television and Social Conflict*, *Film Quarterly*, *The Velvet Light Trap*, and online at *In Media Res*.

Jeffrey P. Jones is Associate Professor for Communication and Theatre Arts at Old Dominion University. He is author of *Entertaining Politics: Political Comedy and Civic Engagement* (2nd ed) and co-editor of *The Essential HBO Reader* and *Satire TV: Politics and Comedy in Post-Network Television*.

Derek Kompare is Assistant Professor in the Department of Cinema-Television at Southern Methodist University. He is the author of *Rerun Nation: How Repeats Invented American Television*.

Elana Levine is Associate Professor in the Department of Journalism and Mass Communication at the University of Wisconsin-Milwaukee. She is the author of *Wallowing in Sex: The New Sexual Culture of 1970s American Television* and co-editor of *Undead TV: Essays on Buffy the Vampire Slayer*.

Amanda D. Lotz is Associate Professor of Communication Studies at the University of Michigan. She is the author of *Redesigning Women: Television after the Network Era* and *The Television Will Be Revolutionized*.

Jonathan Nichols-Pethick is Assistant Professor of Media Studies at DePauw University. He is currently at work on a larger research project about local television in the post-network era. He has also worked as an Associate Producer in the news division of WCSH-TV, the NBC affiliate in Portland, Maine.

Laurie Ouellette is Associate Professor in the Department of Communication Studies at the University of Minnesota, Twin Cities, where she teaches media and cultural studies. She is author of *Viewers Like You? How Public Television Failed the People*, co-author of *Better Living Through Reality TV: Television and Post-Welfare Citizenship,* and co-editor of *Reality TV: Remaking Television Culture*.

Erin Copple Smith is a Ph.D. candidate in the Department of Communication Arts at the University of Wisconsin-Madison. She is currently at work on a project exploring the relationship between increased media conglomeration and media content.

Acknowledgments

Every work must necessarily make exclusions, but in the final drafting of *The Television Will Be Revolutionized* I became steadily obsessed by all I was leaving out. This concern was supported by the sense that my volume was not the only one with key omissions; it seemed every assessment of the "future of television" attended only to the situation of prime-time series. Although I have rarely examined anything but this particular form in my own work, I was well aware of the partial future being imagined and the inadequacy of the situation of prime-time series as predictive of any other television program form.

As I made mental lists of the program forms and contexts unconsidered, I recognized that it would take decades to understand their intricacies adequately enough to assess the consequences of the post-network era upon them. Fortunately, the field has grown considerably, and ready experts existed with the background to offer such consideration. I take great pride in the roster of authors collected here. In nearly all cases (my own chapter on news a primary exception), the authors included reflect on a form of television with which they have established familiarity. Yet the work that follows does not retread their existing writing, but in many cases revisits an area of expertise with new focus. In prompting the chapters, I shared my own concern that the nuances of non-prime-time series programming were being left out of future visions and assessments of the medium, and the following pages offer their responses.

I am consequently greatly indebted to the smart and thoughtful contributors who accepted my challenge, and perhaps micro-management, and delivered detailed assessments of objects of analysis that were shifting as they wrote. All managed the complexity of writing about the present in a manner that will provide lasting relevance and found a way to condense subjects worthy of a monograph into a mere chapter. I am impressed beyond words, and even more grateful.

Introduction

Amanda D. Lotz

By the early 2000s, it was difficult to be unaware of the massive changes overhauling television in the United States and many parts of the world. Evidence of changes surrounded you no matter what your relationship with the medium. Young people and commuters could be seen watching "television" on tiny screens that were notably smaller than the Walkmans that they carried a decade or so earlier, and the functions of personal technology rapidly converged until a single device could be used to make a phone call, send an email, listen to music, or watch videos. As audience members, we'd become accustomed to websites plastered on all types of goods and media just in time for our television networks to begin calling for to us to go online to see "more" of our favorite shows. The nation's television and technology writers repeatedly proclaimed that we were on the verge of the "end of television," and it was difficult to distinguish these apparent threats from something called the "digital transition" that would supposedly render our working analog sets useless. It was entirely unclear whether these shifts were related and who or what might be causing them, but it was obvious that television was changing.

"Television" evokes many different associations for people. For some, it simply brings to mind a technology with a screen: a device that can be used for many purposes, including gaming, watching DVDs, or catching the broadcast signals made to stream through it. To others, "television" might first conjure up the content commonly viewed, so that the term television leads to thinking of the programs that one most prefers to watch—whether a game show, a particular comedy, or maybe just the experience of being in a room with the set on. Indeed, popular understandings of television encompass the device, its content, and much more.

In addition to thinking about television as a technology or as programming, during our more than 60-year history with television, the medium also has been very much defined by its schedule and particular patterns of use that developed in response. News programs, for example, air at specific times throughout the day that relate to daily patterns of work and leisure schedules. Morning hours are filled with talk and "service" programs typically targeted to those who do not work outside of the home and are most likely in the audience at these hours. Evening "prime time" offers programs to an expanded family audience in early hours and steadily develops a more adult focus throughout the night. Particular features of television, such as the patterns of the daily schedule and the experience of waiting for something to "be on," also specifically define our experience with the medium.

During the first phase of television's institutional history—a period spanning from the mid-1950s through the mid-1980s that I call the "network era"—the industrial, technological, and economic norms of television provided most viewers with only a handful of channels that shared this common organization of a programming day. We are now accustomed to tens, if not hundreds, of channels designed to address distinctive audiences or program needs. Some channels offer sports, news, or weather around the clock, while others feature programs specifically targeting children, young women, or teens, no matter the hour of the day. But in the network era, such particular address of audiences or their needs was largely accomplished through the networks' daily schedule construction. Women and children tended to be more likely to view throughout hours in the morning and afternoon, and the networks accordingly devised their programs to match the tastes of the audiences most valued by their advertisers. Throughout history, different forms of television developed with particular industrial features and audience norms based on when networks scheduled them. For example, we associate certain types of talk shows with morning television, newscasts with certain hours of the day, and other forms such as game shows with afternoon and early evening hours. Although the time of day may be the most common way for casual viewers to classify programs, these various program forms also operate with specific budgets, audience targets, and particular value for advertisers that distinguish them in significant and important ways.

"Prime time" is probably the part of the programming day and type of programming with which viewers are most familiar. Even the most casual viewer probably can classify prime time as the period each evening from 8:00 to 11:00 p.m. (EST, MST, PST; 7:00–10:00 CST) when the national broadcast networks typically provide high-budget series for their affiliates.[1] These evening hours commonly gather the most viewers, and the networks consequently design and select shows with a breadth and inclusiveness appropriate to attract a wide array of people. It isn't just chance that programs with the highest production costs air at times that they might reach the largest audience. These shows provide the greatest opportunity for advertisers to reach the most people with a single commercial; this reach of their message makes advertisers more willing to pay higher fees in order to have their commercial messages included during prime-time shows.

Beyond what might be obvious to the casual viewer, though, many typical industrial practices also distinguish prime-time programming irrespective of what time it airs. In addition to the basic economic factors of high budgets and substantial advertiser support, different rules have governed who could make prime-time programs throughout television history. The networks were prohibited from actually making most of their prime-time shows from the early 1970s through the mid-1990s due to a set of regulations called the financial interest and syndication rules. These rules were gradually phased out and completely eliminated in the mid-1990s. Now, one media conglomerate (a corporation such as NBC Universal, News Corp. or Disney) often owns both the studio that produces a show and the network that airs it. For example, Disney owns ABC Television Studios, the studio that produces shows such as *Gray's Anatomy*, *Lost*, and *Ugly Betty*. These shows all air on the ABC network—or, technically, local stations affiliated with the ABC network—which is also owned by Disney.

The costly budgets of prime-time programs require particular financing arrangements. In fact, most producers of prime-time shows initially lose money because the networks typically pay the studio a license fee that covers only roughly 80 percent of production expenses. The networks do not actually buy the shows they air; rather they pay a fee that allows them to "borrow" each episode a few times. If the show is successful, the studio that makes it then sells the rights to air the show not only to that first US network, but typically also to any number of networks around the globe, and later to local stations, US cable channels, and perhaps directly to viewers on DVD. This may seem an unusual system for financing programming, but this practice, called deficit financing, has dominated US prime-time production since advertisers stopped producing their own shows in the late 1950s and early 1960s.

Prime-time programs typically command the highest budgets on television, which allows these shows a particular "quality" look and provides for certain accepted but arbitrary norms—such as that networks typically only schedule and pay for roughly 22 new episodes of a show per year and that they air a new episode no more than once a week, although often they might repeat an episode they've already played before. These norms, which most viewers probably take for granted as just "how US television is," encompass only a few of the industrial practices that distinguish prime-time programming other than the hours during which it airs. The need for prime-time programming to be commercially viable in international and local markets contributes to the topics, themes, and casting of these shows, as do the budgets. Industrial components such as budgets, audience size, profit models, and schedule location vary considerably for different forms of programming. Many of the adjustments that began changing television in the early years of the twenty-first century affected various program forms in distinct ways and tended to make other forms of television more different than similar.

I begin with this explanation of some of the industrial norms of prime-time series television because it is such a defining type of television and one with which most viewers are familiar. The purpose of this brief example is to highlight the ways in which industrial practices and norms affect the creative output of television. US prime-time television looks the way it does because of the budgets made available, the work schedules required to produce 22 new episodes a year, and the scope of audience required to "succeed." The programming produced for other parts of the day look very different—for example, compare the visual style of a daytime soap opera or a talk show. Much of this variation results from the very different industrial norms supporting the creation of these program forms, particularly that their smaller audiences yield fewer advertising dollars and consequently smaller production budgets, and that they might produce as many as 250 new episodes per year.

The future of television was a hot topic in the mid-2000s as the adoption of new technologies such as DVRs, iPods, and online video and related changes in the medium's industrial practices brought substantial adjustments to its forms and uses. With rare exception, however, attention to television's changes focused only on prime-time series programming. Just about every prognosticator offering a vision of the future of television—those who wrote articles and reports with titles such as "The End of Television as We Know It" or "The Death of Television," and even my own book *The Television Will Be Revolutionized*—focused exclusively upon prime-time series.[2] As a corrective to this narrow focus, the rest of this collection ignores prime-time series television and instead examines forms of television that tell a story about the medium that we were otherwise missing at the beginning of the twenty-first century.

The chapters collected here examine how the changes that the television industry began experiencing most clearly by the mid-1980s have affected other forms of programming such as talk shows, soap operas, news, syndication, children's programs, sports, and made-for-television movies, as well as entities such as local affiliate stations and public television. Most all of these forms broadcast their programming during the day, before prime time. In the case of the exceptions, particularly sports and made-for-television movies, the irregular nature of the programming distinguishes it from the regularity of series both in its experience for viewers and in the industrial practices that support its creation. Although prime-time television does sometimes feature sports matches or games, the "event" nature of sports separates it from the norms of scripted series, and it remains the case that distinctive licensing deals with sports leagues also differentiate this content in meaningful ways. Similarly, made-for-television movies were once commonly found on prime-time schedules, but particular features of this form have led it to experience a very different fate than regular series and make it a crucial topic for consideration.

Importantly, all of these forms rely on norms of production, financing, and viewer habits that distinguish them from the practices common in prime time and often from each other. Although the content of these program forms—the types of stories and narratives common to each—might differ significantly from what is typically found in prime time, in their writing here, the authors are particularly concerned with how specific industrial norms differentiate the forms each considers. Using the terminology of "program forms" rather than "genres" is meant to signal this industrial focus, as genre is a classification more common when emphasizing textual features and norms. A multitude of industrial practices differentiate these forms from one another and from prime-time series. Unlike the case of prime time, none of the forms of programming considered here typically relies on the practice of deficit financing discussed above. Some of them produce new episodes every day of the year (news, local, some talk), while others produce new episodes every weekday (soap operas, some syndication, children's). Each of these forms features a particular balance among the costs involved in making the programs and the revenue possibilities that distinguish them—as do the characteristics of their content.

Even though sizable audiences tune in during prime time, we should be careful of assuming that prime-time series provide the most "important" or culturally significant programming. Historically, prime time gathered much of its cultural cachet from the interconnections among its large audiences, expansive advertising revenues, and considerable budgets. As a result, we are sometimes overly impressed by prime time and consequently disregard the importance of other program forms to the institution of television and its future. Prime-time advertising may bring great riches, but networks also must spend substantial sums to find the programming likely to attract those audiences and often lose money in prime time as a result. In contrast, as Elana Levine notes in chapter 2, soap operas provided three-quarters of network profits during the popular height of the form, or, as Jonathan Nichols-Pethick reports in chapter 8, local news, on average, provides nearly half of the profits for local stations. These figures suggest the rich and complicated perspective of the television industry that comes into view once we look past prime time.

Why Reconsider Various Forms of Television?

As the critical study of television developed in the 1970s, early research often attended to the different kinds of programming that could be found on the set. A book that collected chapters much as this one was quite conventional 30 years ago, as researchers attended specifically to various forms of television and made clear distinctions among news, sports, children's programming, etc. (consider, for example, Kaplan's *Regarding Television*, 1983).[3] These categories certainly have not waned in their usefulness for organizing and explaining the medium; however, a complacency of sorts set in after initial studies of various programming forms were completed. In the last decade, the books that have surveyed assorted forms have attended primarily to aspects of genre and offered textual analysis (such as Edgerton and Rose's *Thinking Outside the Box*, 2005, or Creeber's *The Television Genre Book*, 2001), but assessments of the industrial features that differentiate these program forms have been lacking.[4] Admittedly, the preliminary research distinguishing the features of various types of programming served the field well for some time and did not require regular reassessment because many of the norms remained steadfastly in place and only deviated slightly once norms of practice were established. But the sizable adjustments in operating norms developing over the past twenty years have now made it necessary to revisit the attributes and conventions of these programming forms because of the substantial adjustments occurring throughout all forms of television.

The entirety of television has been experiencing significant industrial changes for almost two decades now despite the nearly exclusive attention of industry journalists and scholars to shifts in prime time. In some cases, new technologies or ways of making and distributing television content led to swift and sizable adjustments in the practices of the medium, but in most cases the change has been slow and gradual. The first set of adjustments began affecting television in the mid-1980s. Technologies such as the VCR and remote control afforded viewers unprecedented levels of control over their television experience and enabled ways of using the medium previously unimagined. These devices helped disrupt the long conventional experience of television as an uncontrollable flow of content that came through the set—a flow determined by network executives and station managers whose decisions about what would be "on" television and when left viewers with no recourse. The second adjustment of the 1980s developed from the rapidly expanding choice available to viewers as cable channels, and eventually additional broadcast networks such as FOX, the WB, UPN, Univision, and the CW, launched and became commonly accessible in living rooms around the country. I call the period from the mid-1980s through the mid-2000s the "multi-channel transition" because of the ways that both viewers and industry workers gradually adjusted to new opportunities largely related to expanded choice of and control over television during these twenty-some years.

As the chapters throughout the book explore, these preliminary adjustments of the multi-channel transition initiated different consequences for various program forms. The advent of cable channels focused by topic or type of audience significantly changed the situation of sports, news, and children's programming by the early 1990s. These program forms had been limited to a few hours each week on the broadcast networks, but quickly ranked among the most successful cable channels and created the opportunity for viewers to engage this programming around the clock, all week long. Such competition from cable channels gradually forced broadcasters to adjust many of their conventional business practices; yet, the still dominant broadcast networks continued to find ways subtly to adjust their businesses. Consequently, the industrial shifts of the multi-channel transition were often less profound for the broadcast networks than in relation to viewers' changing experience during these years.

I name these years a "transition" rather than an "era" because change was constant and new developments kept challenging the old ways of making, distributing, and profiting from television. For the better part of twenty years, the industry succeeded in incrementally adjusting old practices, but, by the early twenty-first century, the technologies and opportunities to create and share video became too preponderant for further "adjustments" and a whole-scale revolution began to take place. It is difficult to pinpoint precisely when emerging industrial practices started to take on characteristics of what I call the "post-network era," in contrast to the long list of adjustments that occurred earlier that might be characterized as part of the multi-channel transition. Different forms of television experienced this shift at varied pace, and the sundry components of production—processes such as making, distributing, and financing television, among others—also developed new practices on varied schedules. For prime-time series programming, the post-network era begins to take shape in the early 2000s. As the chapters that follow indicate, this timeline of change does not affect all program forms in the same way, nor do all forms experience a distinct multi-channel transition or post-network era. Although this periodization works neatly for prime-time series, the authors of the following chapters explain the forces of change particularly relevant to their form.

Some shifts affected all forms of programming, although not necessarily equivalently. The erosion of the network schedule provided the adjustment with the potential to bring the most widespread change. The long-dominant schedule mandated that viewers experience television as a medium offering "linear" content—in other words, as a device that enables you to watch specific programs at specific times. The development and slow deployment of the DVR certainly led many to realize the potential of the coming change of television as a more user-driven, on-demand experience, although the pace of adoption of DVRs led them to operate much more as a perceived than actual threat for much of the early 2000s. DVRs entered the US market in 1999 and came into viewers' homes alongside growing video-on-demand (VOD) capabilities offered by cable providers. While many homes gained the technical capability of watching whatever television content whenever they wanted to, the industry fought this revolutionizing potential and made very little programming accessible on demand, thus slowing change; and almost none of the VOD programming was from outside of prime time (although children's programming is a notable exception).

The next development that fractured the linear norm of television emerged in October 2005, and will—I suspect—be remembered by many as the beginning of the post-network era. The agreement between the ABC broadcast network and Apple to distribute a handful of ABC prime-time programs for viewers to purchase and download over Apple's iTunes system left little doubt about the future of television as an on-demand medium. The actual number of downloads and shift in revenue that resulted were insignificant and unimportant. Rather, it was what the deal and ability to use television in this way signified that provided the importance of this development. The agreement enabled viewers to begin legally experimenting with new ways of viewing television.

Of crucial significance for this book, however, is the fact that the erosion of a linear schedule and the threats of DVRs and VOD posed very different consequences for non-prime-time programming. Many of the program forms considered here served a different function in the lives of viewers than prime-time series, and the industrial changes upending prime-time television consequently yielded different implications for other forms. News, sports, and talk particularly derived value from the immediacy of viewing, which made audiences less likely to defer viewing to a later time. There was little point in watching a news broadcast recorded on Tuesday later in the week, because the nature of some events had likely already changed—some might even have changed within a few hours. Likewise, few fans would delay watching a sporting event if the score was already known. The sheer preponderance of content created in other program forms made them similarly unlikely to be recorded. Talk shows and soap operas produce so much new content weekly that viewers quickly fall behind if they do not maintain daily viewing routines, or at least make an effort to catch up once a week. Certainly, DVR or VOD devices might be used to shift programming from its midday airing to evening hours—but the nature of this programming made it likely that people either made time to view regularly or only viewed the episodes that time permitted. Many of these forms derived some of their use just from being "on." Multi-hour daily formats such as the *Today* show provide an ongoing background that viewers could attend to with limited and sporadic focus. These are formats that people are more likely to turn on when they just want to watch "TV," which contradicts the behavior that control devices were acculturating viewers to adopt and involves making deliberate choices to watch specific programs.

Although the post-network era introduced technologies and opportunities to use television likely to revolutionize prime-time programming, the variation in the institutional norms and viewer uses of other program forms made them unlikely to precisely replicate the adjustments experienced in prime time. Many of the forms considered here functioned crucially in daily rituals of television use—whether of those who habitually turned on the set while preparing for work or school, upon arriving home, or to pass time throughout the day. The significance of ritual use is very different from the active fanship that leads people to set a weekly recording and make time to view a weekly episode of a prime-time show, and, consequently, this distinction must be accounted for in our thinking about the role of different kinds of television in the cultural life of a society and how the post-network era may be changing it.

The organization of the book roughly follows the conventional schedule of the broadcast day and then includes two chapters that transcend the daily experience. The earlier chapters emphasize those program forms in which ritual likely functions most strongly, as in many cases the experience of watching forms such as morning talk, afternoon soap operas, and evening news have been an instrumental component of viewing. The later chapters on sports, made-for-television movies, local affiliate stations, and public broadcasting are less defined by ritual experience, but also distinct from prime-time series. Post-network developments that enable greater viewer choice and control have initiated notable new possibilities for some of these forms of television while threatening the continued existence of others.

Jeffrey Jones leads off the collection with a consideration of talk shows—here broadly defined to include the hours of network self-produced morning shows such *Today* and *Good Morning America*, as well as late-night productions such as the *Tonight Show with Jay Leno*. Despite the diminishing audiences in most other programming forms, talk shows remain popular and continue to thrive among audiences and even return profits. Jones explores how the features of intimacy and familiarity that are particularly pronounced in this form have allowed networks to utilize the interactivity and opportunities for connection provided by new web technologies to expand the reach of talk shows—as well as the network brand. Jones also analyzes how the networks are anchoring broad network-wide competitive strategies to initiatives inaugurated by talk shows.

In chapter 2, Elana Levine examines the conflicting narratives surrounding the declining ratings of soap operas and interrogates how these varied (and sometimes wholly manufactured) explanations have meaningful implications for perceptions of this cultural form and its continued viability. Levine carefully traces the consequences of industrial changes, but also considers the adjustments in the broader socio-cultural sphere to present an array of forces that combine to shift long-established viewing patterns and instigate questions regarding the future of the form. As is the case in so many chapters here, although current industrial and cultural configurations challenge the reliable norms of the past, so too do new opportunities emerge, and Levine explores new strategies being used by soap opera producers to contain production costs and create new revenue streams.

Derek Kompare assesses the broad terrain of syndicated programming in chapter 3. Syndicated programming loosely describes programming sold directly to local affiliate and independent stations and encompasses both first-run programming, such as *Oprah*, *Judge Judy*, and *Jeopardy*, as well as programming that has already aired on a network or cable channel station, such as *Seinfeld*, *Friends*, or *The Shield*. Kompare skillfully explains the industrial shifts that particularly affect this form—most notably the consolidation of distributors—while also considering how characteristics of the experience of syndicated programming have led to stability despite the general uncertainty in the television industry.

In chapter 4, Sarah Banet-Weiser looks at the changing nature of children's television and charts the shifts that call into question the central methods for studying this programming and even its validity as a meaningful categorization. She identifies two major disruptions—first, the creation of dedicated kids' cable channels and, second, the rise of brand culture—as key developments that have altered the nature of what television makes available to the children's audience. Banet-Weiser provokingly suggests that kids' television cannot be disentangled from the multi-mediated, interactive, and branded environment of kids' media culture, a reality that necessitates new approaches to examining children's television.

As afternoon turns to evening, in chapter 5, I explore the state of and prospects for the 6:30 p.m. national nightly newscasts. Outside of prime-time series, the fate of this form has received the most critical attention, and the broadcasters' national newscast has been the subject of repeated obituaries since the end of the network era. This chapter traces the developments of the last 25 years—those both real and apocryphal—in an attempt to provide a chronologically grounded examination of the genuine and imagined threats upon this programming that offers a reflection and record of the events of the day. I then consider some of the attributes of news and the ways in which the technologies of the post-network era create opportunities to redefine the experience of video news.

In chapter 6, Victoria E. Johnson assesses the paradoxical status of sports on television. As she notes, many aspects of sports programming maintain network era norms, particularly that it often draws mass, heterogeneous audiences and is almost always viewed live. At the same time, sports programming is also very much augmented by post-network viewing possibilities such as the personalized microcasting (the extreme of narrowcasting) enabled by online distribution of sports events, the benefits of high-definition and other new technologies that enhance the sports viewing experience, and the breadth of tools for participating in sports fanship made feasible by the integration of television and Internet. Johnson deliberately works through the many distinguishing features of sport on television to indicate the complicated tensions characteristic of this very particular, and culturally central, form of programming.

Made-for-television movies once occupied one-quarter of the networks' prime-time schedules, reports Erin Copple Smith in chapter 7, but by the early twenty-first century the form was nearly absent from broadcast schedules. The fate of the made-for-television movie has been quite different than that of prime-time series, as various aspects of the industrial transformation of the television industry have led to its near-complete disappearance from broadcast networks in recent years. Contrary to popular perception, however, the form lives on and thrives on cable channels and in international markets. Copple Smith charts how the textual characteristics that threatened the made-for-television movie in some markets have proved its greatest assets in others.

The final two chapters consider somewhat broader topics. First, in chapter 8, Jonathan Nichols-Pethick assesses the state of local affiliates with particular attention to their local news productions. The local affiliate stations exist paradoxically in the post-network era as technological development has made them irrelevant, yet those stations owned and operated by the networks provide their most imperative source of revenue. Through detailed explanation of the economic, regulatory, and technological changes of the last two decades, Nichols-Pethick illustrates the many contradictions faced by local stations that make them simultaneously essential and vulnerable and could inaugurate for them a new phase of operation.

In the final chapter, Laurie Ouellette examines public broadcasting. While this is admittedly more a network or even a mandate for media industry operation than a program form, too often books such as this exclude consideration of public broadcasting from their scope. Although the chapter appears to focus on a topic of more macro-scope than a single program form, the distinctive industrial norms that differentiate public television from commercial television provide considerable similarity with the previous chapters. Here Ouellette assesses the strategies that PBS has used to respond to the adjustments of the post-network era. She explores its loss of place that results from its fading program distinctiveness in a television landscape that now includes the offerings of cable and the loss of legitimacy that emerges from the ways "neoliberalism" questions government involvement in the enterprise at its core.

One advantage of bringing together assessments of various program forms in a single volume is that both continuities and surprising gaps appear in multiple chapters. The O. J. Simpson trial of the mid-1990s and the way ample coverage of the event disrupted daily schedules for some forms emerge in the discussion of both the current state of soap operas and network newscasts. Similarly, multiple authors reference the consequences that conglomeration of the media industry has had upon their program forms. Notably, these consequences are not uniform or consistent, but, rather, the forms have commonly had production conventions altered as a result of shifts in industry organization.

Readers may identify other chapters of interest missing from this edition. Perhaps because of the attention to the industrial practices that differentiate program forms—instead of textual or narrative analysis of genres—the peculiarity of the US context makes this appear a particularly US-centric work. Although many other similarly industrialized countries experienced comparable shifts in their television industries and many of the forms discussed here circulate across national boundaries, different countries often feature different production norms that may allow minimal transnational application. Even though some of the programming addressed here—particularly the made-for-television movie—came to require international circulation to be financially viable in the United States, it also bears heavily the imprint of the context of its creation in a manner that makes it culturally specific in an industrial sense, regardless of its transnational circulation or the production of similar forms in other countries. The nearly entirely free-market, commercial structure of the US television industry weathered the adjustments of the multi-channel transition and post-network era in a manner distinct from, although not unrelated to, the experience of those shifts in national television industries that originated in and maintained some semblance of a public television mandate alongside encroaching commercial structures.

In addition to the limits of its US-centricity, the collection has perhaps clumsily combined the consideration of program forms that might have been addressed in multiple chapters—as in the case of the broad syndication chapter or in addressing somewhat distinctive forms and formats in the single "talk" chapter. Other important components of US television—such as Spanish-language television—are left unconsidered because of its complicated history and breadth. Despite these inevitable errors of omission, the chapters assembled here do contribute significant work toward developing a more all-encompassing understanding of the consequences of changing industrial norms on television as a whole.

One of the key issues suggested by this collection of chapters is the inadequacy of most claims made about "television" these days. The medium certainly always has been varied—as most of the forms considered here have roots in television's first years. Still, the centrality of the networks—which long only numbered three—and the dominance of shared industrial practices made it possible to conceive of television as a somewhat unified subject. The adjustments throughout the business and the variation in practices that now can be found within and among different forms make broad pontifications about the television of today or the future increasingly meaningless. Moreover, those who use prime time and its scripted series as the bellwether for the medium overlook the vital and necessary economic role played by other program forms that are also fundamental to television.

Notes

1 On Sundays, prime time has traditionally spanned from 7:00 to 11:00 p.m.
2 IBM Business Consulting Services, "The End of Television as We Know It," 27 March 2006, http://www1.ibm.com/services/us/index.wss/ibvstudy/imc/a1023172?cntxt=a1000062&re=endoftv, accessed 19 April 2006; Adam L. Penenberg, "The Death of Television," *Slate.com*,

17 October 2005, http://www.slate.com/toolbar.aspx?action=print&id=2128201, accessed 20 October 2005; Amanda D. Lotz, *The Television Will Be Revolutionized* (New York: New York University Press, 2007).

3 E. Ann Kaplan, ed., *Regarding Television: Critical Approaches—An Anthology* (Fredrick, MD: University Publications of America, 1983).

4 Gary R. Edgerton and Brian G. Rose, *Thinking Outside the Box: A Contemporary Television Genre Reader* (Lexington: University Press of Kentucky, 2005); Glen Creeber, *The Television Genre Book* (London: British Film Institute, 2001).

Chapter 1

I Want My Talk TV

Network Talk Shows in a Digital Universe

Jeffrey P. Jones

The persistent decline in network audience ratings has been a central factor in the institutional restructuring of broadcast television networks in the post-network era. While cable and other technological innovations provided audiences with viewing alternatives during the multi-channel transition, the vast array of digital technologies now available to consumers—and the migration of viewership across these new media platforms—has contributed further to the networks' ratings decline. From 1980 to 2005, network audience share fell from 90 percent of those watching television to 46 percent.[1] During the same period, network news lost half its audience.[2] The continued ratings achievement of network talk shows is all the more remarkable when viewed in this context.

Talk shows have been a successful programming form throughout the history of television, typically comprising each network's morning and late-night dayparts and often contributing significantly to brand identity. Audience loyalty to such programming has been remarkably consistent from the earliest days of television to the present day. Late-night talk show viewership actually increased by 40 percent from 1995 to 2005, and the form continues to have a dedicated fan base.[3] Network morning talk shows are even so immune to ratings decline that industry observers have historically described them as "bullet proof."[4] It is estimated that morning shows produce over 1 billion dollars a year in advertising revenue, double what the nightly newscasts generate. In fact, these shows often help fund most of the networks' news operations.[5] Furthermore, because talk shows are comparatively cheap to produce, profit margins tend to be quite high. The *Today* show, for instance, reportedly earned $250 million in profit from $500 million in revenues in 2006.[6] And, for two of the three networks, the morning talk show franchises have even *expanded* their programming hours in recent years. In short, while the network ratings decline has forced an industrial reconfiguration by the broadcast networks to address eroding audience share, talk shows as a programming form have not contributed to such need for change.

Talk shows are nevertheless important elements in the networks' digital strategies and part of their overall organizational alterations in post-network architecture. The reasons for this include the particular features of talk show content and form as well as opportunities presented by new distribution technologies. Talk shows provide popular content and recognizable faces (often closely associated with the network brand) that can be used to hail viewers across crowded and dispersed media platforms. And because talk shows are created in specific segments, they have a natural structural advantage for easy distribution, downloading, and viewing as short snippets in a variety of settings and through multiple technologies (from web and mobile devices to out-of-home viewing screens). As such, the content and form of talk shows are conducive to addressing these changes in distribution, as opposed to the longer 30- and 60-minute long-form programming found in prime time.

But this process is intensive as well as extensive. That is to say, networks are also using digital technologies to craft different relationships with both audiences and advertisers. Networks have learned to exploit audience desire to interact and engage with programming, including sharing and discussing topics in online communities. The networks have recognized that talk shows have the potential to play an expanded role in viewers' lives, more so than when such programming was available simply as a morning or late-night ritual (that is, what someone watches during periods of attenuated brain power or while doing something else). Instead, talk shows can become sites for extended viewer engagement with the network brand, as well as the means for engagement with other viewers. Furthermore, as audiences engage talk show content across platforms, these new technologies allow networks to monitor and aggregate much more information about their viewers—their tastes, interests, lifestyle, and patterns of behavior—than that typically available through the ratings data that have dominated the industry for decades. Networks are increasingly learning not only to exploit this information with advertisers, but also to use it to inform programming decisions.

This chapter explores these and other issues related to the role of talk shows in post-network realignment. I begin by discussing the talk show as programming form: its textual features and traditional role in network performance. The discussion then turns to a broader examination of post-network changes in the relationship between networks and audiences before offering an analysis of morning and late-night dayparts and how they are being used in network realignment strategies. But, first, it is important to define what exactly is under investigation here. Talk shows comprise an enormous swath of programming available to viewers across all dayparts and across both network and cable channels. The shows examined here, however, include only programs that are self-produced by the networks and fall primarily within the broadcast networks' morning and late-night dayparts. While some popular syndicated fare (*The Oprah Winfrey Show*, *Live with Regis and Kelly*, *The Ellen DeGeneres Show*) and cable talk shows (*The O'Reilly Factor*, *Countdown with Keith Olbermann*) are connected to the broadcast networks either through distribution or appearing on cable channels owned by the same corporate parent, they do not have a similar central relationship to the network as those shows owned and produced by the networks for airing on broadcast television. The latter (as opposed to talk shows acquired from independent production companies) permit the networks much more leeway in how such programming can be used (including its editing and reuse, its franchising and off-shoots, and so on). Therefore, the focus solely on network talk shows allows for an investigation into how the networks are leveraging this form of popular and profitable content in their overall network-wide digital realignment strategies.

Network Talk Shows as Successful Dayparts

Network talk shows fall primarily into two dayparts—morning and late night—and have dominated those programming hours almost from the medium's beginning. Among weekday morning talk shows are *Today* on NBC, *Good Morning America* and *The View* on ABC, and *The Early Show* on CBS. The *Today* show has been the ratings leader for most of the program's history, including ranking continuously as the number one morning show since 1995. The program has been so successful that it expanded its original format of two hours to three hours in 2000 and then to four hours in 2007, using a slightly different format for the last hour. *Good Morning America*, another popular though slightly less profitable program, also expanded to a third hour (as *Good Morning America Now*) in 2007, although this hour is only available via cable, online and mobile media. As a genre, morning talk shows also include the long-running Sunday morning public affairs talk shows such as *Meet the Press* (NBC), *Face the Nation* (CBS), and *This Week with George Stephanopoulos* (ABC). Although the Sunday talk shows continue to garner strong viewership numbers, their specialized status and restricted broader appeal (not to mention their older demographic) has somewhat limited their value as programming that the networks can foreground in their digital strategies (and are therefore tangential to the discussions here).

Weekday morning talk shows are organizationally related to network news divisions (*The View* as exception), but are widely considered both institutionally and by audiences as "talk shows" more than "news" programs. These shows perfected the blend of information and entertainment into a popular format long before critics began using the pejorative term "infotainment" to describe the blurring of boundaries between the two. In fact, the prototype for morning talk has been the *Today* show, which first aired on NBC in 1952. In its first broadcast, host Dave Garroway explained how the program would bring viewers news, but also information about music, art, science, sports, and "all fields of endeavor we think we'll be able to inform you better about …."[7] By contrast, the content that largely fills morning talk today is focused on topics that the networks believe will best appeal to the women viewers who comprise the vast majority of the audience.[8] The programs open with a news update and incorporate additional news segments and newsmaker interviews interspersed throughout the show. The bulk of programming, however, tends toward lighter infotainment fare, including segments on topics such as parenting and family, fashion and beauty, food and cooking, relationships and sex, homemaking and gardening, health, travel, personal finance, weather, celebrities, and musical performances. The historical contrast is instructive in that the focus on women as primary viewing audience and the gendered grouping of topics have become two of the central areas through which morning talk has been exploited in the networks' digital strategies (as discussed below).

The infotainment approach to morning talk subjects has traditionally been designed with the "homemaker" in mind, recognizing that lighter content requires less concentration in viewers who are typically occupied by morning breakfast and family mobilization routines. Also central to talk show appeal has been the aura of intimacy and friendliness (between the cast members and with viewers) that the morning talk team is capable of creating. The casts are most successful when projecting themselves as likeable companions, even a family, such as the *Today* show's co-anchors Meredith Viera as bubbly sister and Matt Lauer as steadfast brother, with Al Roker and Willard Scott as wacky but loveable uncles. Katie Couric (*Today*) and Diane Sawyer (*GMA*) are also two recent hosts that have been enormously successful in crafting a pitch-perfect relationship with viewers through their on-screen personas. Finally, morning talk shows also try to steer clear of controversial topics, although breaking headlines or long-running news events can sometimes make such topics unavoidable. When ratings began to sag for all morning shows in spring of 2007, networks and advertisers questioned whether network audience drift had finally reached the morning daypart or whether (women) viewers were simply experiencing fatigue with stories such as the wars in Iraq and Afghanistan.[9] Later that year, when the *Today* show needed to cover the increased public awareness and concern over the gloomy issue of global warming and climate change, it was able to get a big bounce in ratings by sending its crews to the north and south poles and the equator (titling the extended coverage "Ends of the Earth"), thereby transforming a potentially downbeat story with catastrophic implications into a fun travelogue piece, later available for purchase on DVD for $29.95![10]

As with the morning shows, the late-night daypart has historically been dominated by entertainment talk shows. Among late night network talk shows are *The Tonight Show with Jay Leno* and *Late Night with Conan O'Brien* (NBC), *The Late Show with David Letterman* and *The Late, Late Show with Craig Ferguson* (CBS), and *Jimmy Kimmel Live!* (ABC). These shows too are largely driven by the personality, talents, and charisma of their hosts (recognized by their name's placement in the show's title). And here too, although to a lesser degree, the shows attempt to create a feel-good repartee (in this instance, of male "buddies") between host and co-host or bandleader. Late-night talk is also a highly segmented programming form, including a comedic monologue, comedy sketches, video vignettes, musical performances, and celebrity guest interviews. Most comedian hosts have perfected some sort of signature comedy piece that is prominently and repeatedly offered, such as Letterman's "Top Ten List" and "Stupid Pet Tricks" or Leno's "Jaywalking." Similar to morning talk, late-night talk is designed for ritualistic viewing. Although the comedians might engage in spectacle or perhaps occasional shock comedy, they rarely offer up much in the way of unscripted or controversial material. Although ABC flirted with such a format for several years with the political entertainment talk show *Politically Incorrect with Bill Maher* (1997–2002), most late-night talk is designed to amuse rather than bemuse the viewer.

Both morning and late-night network talk shows, therefore, are long-running programming forms that have proven highly popular with audiences for many of the reasons just described. And, as noted, neither daypart has experienced the levels of audience erosion evident during prime time. Given this cheap and profitable form with identifiable stars, segmented light-entertainment content, and a loyal audience base that shares a ritual relationship to the form, the networks have attempted to exploit talk programming in their efforts toward industrial reconfiguration. Primary to such moves is the broader reformulation of the relationship between networks and audiences. The discussion now turns to an examination of this reformulation and, in particular, how this altered relationship has direct bearing on the role that talk shows are being asked to play in the post-network industrial model.

Changing the Audience—Network Relationship

Arguably, the most important change in post-network television is the transformed relationship between audiences and broadcast networks—a transformation that in many ways redefines traditional definitions of both. Industry executives realize that, while the value in networks lies in the creation of successful content, as businesses they are being forced to transform themselves from content companies to audience companies.[11] This reformulation has several meanings. The first relates to the distribution of content. Television networks can no longer simply rely on the production of quality content to assure their financial success as they could—at least to some degree—in the oligopolistic network era when audiences had fewer viewing choices and little control over what they watched. Instead, as audiences have become widely dispersed across an array of media platforms—DVRs, DVDs, a variety of web locations, mobile devices (including mobile phones and iPods), and others—television networks must go where audiences are located instead of simply counting on audiences coming to them. Networks must be very intentional about how they distribute their content outside of the traditional network-affiliate relationship precisely because viewers have so many choices, including the choice not to watch television altogether. As Jeff Gaspin, president of NBC Universal Television Group, describes it, "The shift from programmer to consumer controlling program choices is the biggest change in the media business in the past 25 or 30 years."[12]

In attempting to go where audiences are located, the networks have employed several strategies. One has been to own and control the means of distribution either by buying up important Internet and cable properties (such as News Corp.'s purchase of MySpace, CBS's purchase of CNET, and NBC Universal's acquisition of iVillage) or by creating their own portals to which consumers must come for free ad-supported content (such as NBCU and News Corp.'s partnership to create Hulu.com). So, for example (as discussed at length below), NBCU decided to build on its top-ranked *Today* show franchise and existing cable channel Bravo (both highly popular with women viewers) to reach an even larger female demographic by extending the network's reach across media platforms. Hence, in addition to its purchase of the web community iVillage, it also acquired the cable channel Oxygen, TheKnot.com (a wedding planning website), and Pop Sugar (an online community), while also forming a content sharing partnership with BlogHer, a female-oriented community of 2,200 individual blogs. The network combined these media assets into a Women@NBCU division to coordinate cross-media content-sharing, collaboration, promotion, and advertising. While there is little evidence yet that such acquisitions are enhancing network value, it is nevertheless important to highlight the network's belief that its popular and highly profitable talk show franchise was the place to start in crafting and building a plan for broader distributive reach and cross-platform synergy.

An alternate approach has been to create strategic partnerships for the sharing and distribution of video content with a variety of new media companies (such as iTunes, YouTube, AOL, Microsoft, Yahoo!, Fancast.com, and Joost), mobile phone carriers, and others. Fancast, in particular, allows Comcast subscribers to view content across screens or on the screen of their choosing.[13] After a failed attempt to obtain proprietary control of content distribution through its own web portal (an initiative called Innertube, which one CBS executive later ridiculed as CBS.com/nobodycomeshere), CBS quickly learned that it made more sense to let its content flow freely across the web through outlets not under its control but through which it nevertheless received remuneration. As CBS Interactive president Quincy Smith explains, "you don't win by telling everybody where they should go, you win by being where everyone else is."[14] As a result of developing what the network now calls its CBS Interactive Audience Network, it extended its "effective reach" from 13 percent of the web to 92 percent.[15] So, as we will see in the discussions that follow, the short segmented form that largely comprises talk programming such as David Letterman's monologues and "Top Ten Lists" becomes the perfect content form for such broad-based distribution in outlets such as YouTube. Furthermore, as viewers virally distribute popular content such as Jay Leno's comedy, networks have less need to try to contain or restrict its distribution. If viewers are going to YouTube to watch pirated content owned by the networks, for instance, recent evidence suggests that, rather than force YouTube to remove the content, media companies such as CBS tend to leave the material online (about 90 percent of the time) and simply share in the advertising revenue aired alongside it.[16] It is worth noting that talk shows previously had little aftermarket value, that is, the ability to be used outside of their original airing (as opposed to syndication for prime-time programming). The networks are now realizing that distribution and viewing in contexts outside of broadcasting is producing new revenue streams.

With either of these strategies, though, the need to redress the increasing tendency for networks to receive "digital pennies for analog dollars," as NBCU CEO Jeff Zucker describes it, is a top priority.[17] What Zucker refers to, seemingly, is that the digital avenues of content distribution have yet to demonstrate a level of revenue comparable to that provided by the older broadcaster–advertiser relationship. Television programming still costs considerable amounts to produce, but can be paid for by traditional advertising dollars. The rate that advertisers are willing to pay for content that moves through digital distribution channels, however, amounts to much less (or in Zucker's language, "pennies"). To remain profitable the networks must learn to drive that content through many different media and devices. While Zucker is most likely referring to expensive prime-time programming (such as *Heroes*), we can nevertheless see the relevance to talk programming as well. The cost of production for a show such as *The View*, for instance, can be recouped if ABC allows for the free flow of its clips (such as appearances by Barack and Michelle Obama) across numerous advertiser-supported platforms and tallies the payments earned. And, as already noted, such segmented fare can produce new revenue streams that can help offset revenue lost in prime time.

The second meaning of the transformation from a content company to an audience company relates to the way in which networks realize the importance of "community" as a means of establishing, extending, and/or reformulating relationships with viewers. As interactive media in the digital era has "liberated" audiences to be more than simply consumers of content, there is an increasing desire for avenues of engagement and participation with all forms of media. Creating opportunities for audience interaction with network content, therefore, is a key network strategy, or as Jonathan Lees, president of CBS Television Stations Digital Media Group, explains it, "a necessity." He notes, "People want to be heard, people want to be recognized."[18] Furthermore, they want to feel they are part of something. As Beth Comstock, NBCU's president of digital media and market development, defined it, "In the digital age, community is all about gathering people with shared interests and giving them a platform to interact with each other, to engage in relevant content, and to create something new."[19] A goal of transforming its relationship with audiences led to NBCU's purchase of iVillage. CEO Bob Wright explains, iVillage "knows how to create and sustain communities; we know how to create great content."[20]

The inclusion of community-building features on websites such as iVillage has become a means for transforming the traditional one-way communication of television to a more interactive experience, a development that arguably has particular significance for a program form such as talk shows. Websites allow for viewer participation in ways that can increase interest in the brand, foster brand loyalty, build relationships and feelings of community among users, generate free content, and provide insight into viewer behavior and interests through interactive features such as online surveys and the ability to share content with fellow users. Viewers can also often engage with producers and stars, as possible through Meredith Viera's blog for the *Today* show. Of course, "community" may mean little more than developing an enhanced relationship between viewers and the brand. As Wright noted about NBC's purchase of iVillage, "After the integration is completed, we'll be able to bring video to bear to better take *advantage* of the community that iVillage has built up."[21] Similarly, CBS is attempting to build upon and exploit the community-oriented features of its recently acquired streaming music site Last.fm. As one press report described it, Last.fm "allows users with similar habits to find each other and recommend brands in a way that goes beyond the more straightforward 'friends' mechanism of MySpace."[22] The network has focused on developing a video player that will allow for "content sharing features and social-viewing rooms that let friends watch and discuss content together."[23] Social networking sites, therefore, provide another opportunity for the networks to exploit the desire for connection, community, participation, sharing, and engagement that audiences seemingly desire by linking them in newly creative ways to their brands. In sum, the overall effect is that the networks are attempting to create a new relationship that will lead viewers to new forms of engagement with network content—in myriad ways, across numerous platforms, and with greater frequency (as we will see below with the *Today* show and iVillage).

The networks' ability to monitor and measure their audiences' behaviors and interests (not just demographics) with much more specificity than previously possible has evolved in this digital environment. The third aspect of the transformation into an audience-focused company, then, is the networks' realization that such precise knowledge of audiences has enormous value in its own right. Subsequent efforts to capitalize on their access to this information are now being foregrounded in their digital strategies. "Traditionally, the focus has been on the outbound message," says Tim Armstrong, vice president of ad sales at Google. "But we think the information coming back in is as important or more important than the messages going out. For years, demographics has been a religion among advertisers because it was the only information they had. Now they're realizing there's more out there."[24] So too have the networks, as can be seen in the actions of both CBS and NBC described above. A *Daily Variety* report on CBS's intentions in acquiring Last.fm argued that CBS is "less concerned in using the web to promote its shows or with developing a parallel television network online ... as it is with amassing as much information as possible about its viewers' tastes."[25] CBS's Smith concurred, even refining his proclamation of change from a "content company" to an "audience company" by more precisely noting, "We're evolving from a content company to a media-profiling company."[26] One of the top reasons that CBS acquired Last.fm was for the technology the company employs. CBS is interested in modifying the social networking technology for video content, thereby allowing users to compare their profiles and viewing preferences and comment on programming. What such a product would provide, according to CBS, is a wealth of new audience data and metrics that could produce more behavior targeting in advertising.[27] So, for instance, as users comment on Craig Ferguson's monologue or share Letterman's "Top Ten Lists," technology that would allow the network to monitor such behaviors might glean additional insight into viewers (such as who they are, what ads they are clicking, what other websites they've visited, and so on). It is unclear whether CBS's attempt to exploit social networking technology such as Last.fm will ever come to fruition. Of interest here, though, is the recognition that insight into how audiences use and relate to content provided by new media technologies can be of new value to the network.

Furthermore, talk programming content offers a variety of avenues that can provide greater insight into viewers' identity, lifestyle, and habits. *Advertising Age* reported that NBCU is trying to offer its advertising clients more information than pure demographic buys. The magazine reported that, "after identifying Bravo's audience of tech-savvy, big-spending 18-to-34-year-old consumers as 'affluencers' last year," NBC's marketing executive "has been working with each of her network's research teams to identify key behavior patterns and spending habits. Recent NBCU acquisition Oxygen, for example, is being rebuilt around 'Generation O,' or 'trenders, spenders and recommenders,' with similar assignations for each of NBCU's digital and on-air brands."[28] Although Bravo does not air much in the way of talk programming,[29] the example is nonetheless instructive because we can see how the particular features of talk programming described earlier (especially morning shows' focus on lifestyle segments such as fashion, relationships, travel, health, and so on) might provide valuable insight into viewers' behaviors as they engage such topics in digital platforms such as the Internet, where shopping and consumption can be part of the online experience.

NBC is also banking on the power of new metrics as a driving force for profit in its acquisitions of new media properties. By combining content from Oxygen

Media, Bravo Media, iVillage, and other properties with the *Today* show, NBC has created a "virtual network" for advertisers trying to reach women. The goal, says Mike Pilot, president of advertising at NBCU, is to "bring the entirety of the NBC Universal portfolio of assets to the advertiser in one package." But, as Pilot emphasizes, such institutional arrangements are not just a means for doing the same old thing only better. Rather, he stresses that the network's new understanding of audiences provides unexplored opportunities for the creation of revenue. "We've got to resist allowing these new ideas to be cranked through the same old CPM [cost per thousand] machine. We've got to demonstrate new value and we've got to realize that new value."[30]

In sum, then, this altered relationship between audiences and networks redefines them both. The network is no longer a company whose primary function is to produce content but, rather, one that is capable of locating very particular audiences across an array of media and engaging them with a set of experiences and relationships to capture and hold their attention while measuring their behaviors. Video content, therefore, might be only one in a variety of such experiences/relationships that the network provides. The network may be doing little more than facilitating interactions around its brand in its efforts to draw and sustain audience attention. Similarly, audiences are redefined in this process as well. Audiences are not just consumers of content, segmentable by demographic categories, who will watch whatever is the least objectionable programming on television at any given moment. Rather, they are people with varied sets of behaviors, interests, needs, and desires who enjoy greater ability to choose and control what they watch. They are people who will interact with content and, indeed, desire to be engaged with media materials in complex ways. Thus, through their interactions, they extend the network's brand by having an intensified relationship with it.

These characteristics of a reformulated relationship between audiences and networks are what NBCU's Comstock had in mind when she outlined the "3 C's of Success" in the digital era.[31] She argued that broadcasters must focus on *context*, *community*, and *content* if they hope to be successful in navigating the new media landscape. "The way consumers use media," she contends, "is different at different times of the day and in different situations." Television executives must understand this and make television "relevant" in the context of viewers' lives. Will people watch talk shows on a two-inch screen? It depends on the context, for instance, the situational contrast between receiving Conan O'Brien's monologue on a train ride home from work versus watching the entire show on an iPod while sitting in one's living room. Community is also an important means of making television relevant, especially in a program form such as the talk show that relies heavily on intimacy in presentation. And, as she notes, both of these are built on quality content—what she calls the "great differentiator" in the era of infinite content—something that networks have been historically successful in providing. This, she argues, is NBCU's strategy: "to develop and extend great content, to make sure it's relevant and put into the context of our viewers' lives, and to build efforts around communities of users and their interests."[32]

While talk shows may not generate the buzz that has accompanied the availability of prime-time programming in new distribution platforms, they nevertheless fulfill the criterion of success that Comstock suggests is needed in the digital era—and perhaps even more so. Popular talk show segments can be easily distributed to dispersed audiences in a variety of contexts. Both the humorous and lifestyle aspects of late-night and morning talk, respectively, provide opportunities for engagement and community. And the talk franchises that the networks maintain in both dayparts are valuable properties that can grow and extend the brand. While talk shows continue to play an important role in network profitability, they are also one area through which the current redefinition of the network–audience relationship is occurring. The discussion thus turns now to one prominent example of this process currently underway at NBC's *Today* show.

From Talk "Show" to Community and Virtual Network

In the transition from a traditional broadcast format to a program integrated into the 24/7 digital environment, *Today* has grown from a show that aired for two hours and then signed off until the next morning to one with a continual and enticingly deep presence on the web. This is to say, what at one time might have been considered ephemeral programming now becomes the basis for a reservoir of information that can be cataloged, archived, linked, and accessed at any time of the day or the week. An analysis of the *Today* show's webpage conducted in October 2007 indicated that it included a netcast of the show's first hour for each day of the week; a gallery of 25 various video segments from the most recent shows; a link to 149 video clips from each day's broadcast for the previous week (what amounts to each show broken down into its segmented parts); and a frame of 19 tabs for links to various topic areas regularly covered in the show's programming (health, beauty, weddings, relationships, and so on). In total, the website contained 272 video clips across the entire site, as well as an enormous array of print articles within these various subject areas.

But, beyond serving as an archive or library of items, the web also integrates formerly distinct behaviors that might exist outside media consumption routines. So not only will a visitor watch television, but she also has the opportunity to engage in shopping (from advertising to downloads of *Today* show musical performances via iTunes), talking and relating (discussion boards and blogs), and reading (articles). The relationship to the program, therefore, becomes much more complex

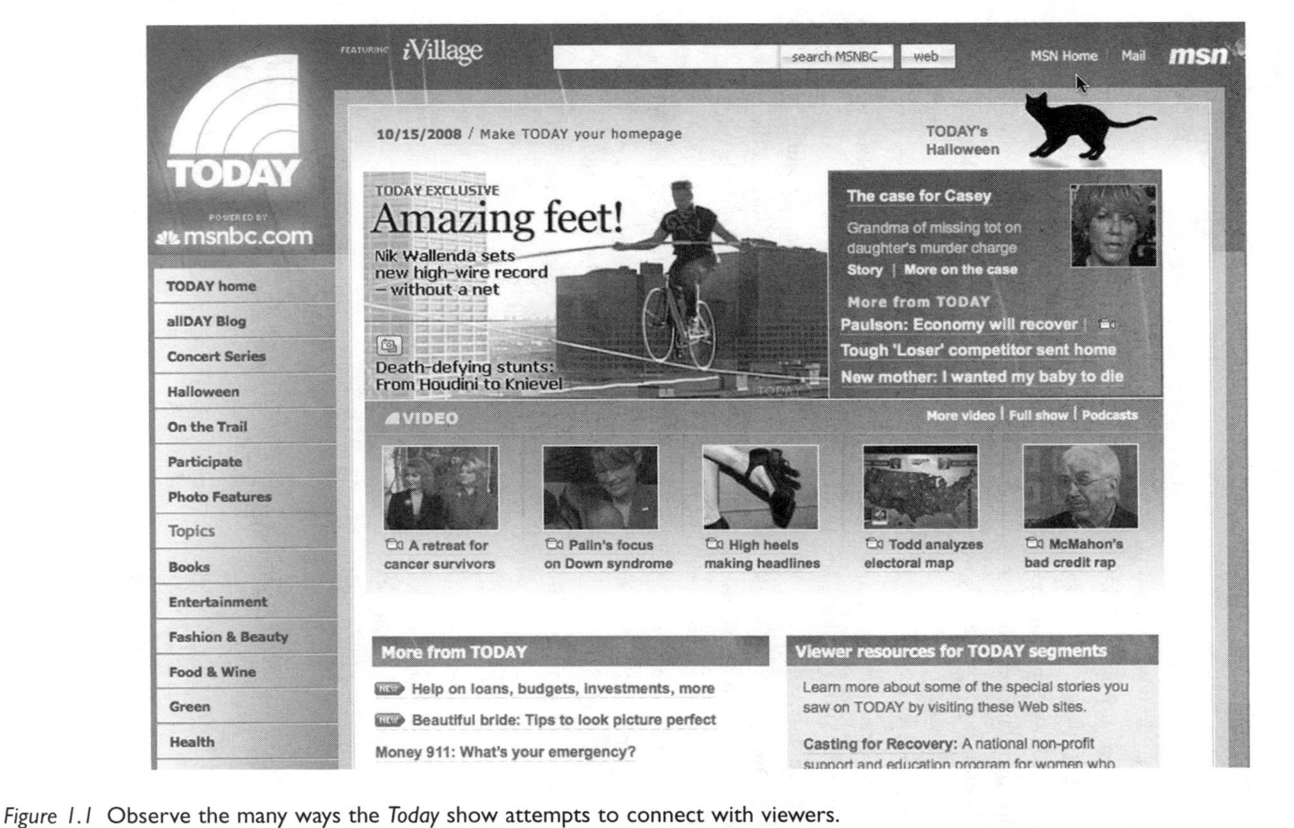

Figure 1.1 Observe the many ways the *Today* show attempts to connect with viewers.

Source: http://today.msnbc.msn.com/ (accessed October 15, 2008).

and interactive than was previously the case. Viewers can participate in surveys, upload photos, search for recipes featured on the show, submit questions, share knowledge and experiences, send on-air birthday wishes to other viewers, and more. The "relationship" with the hosts and producers is extended further through the program's allDAY blog, where the cast and producers write features about their activities, experiences, thoughts, feelings, and the like in producing the show. Finally, the webpage provides opportunities for visitors to subscribe to video or audio podcasts, mobile alerts, and even become Facebook "friends."

Whether any or all of this adds up to feelings of "community"—however one chooses to define the term—in *Today* show viewers is debatable. There is no question, however, that the potential for engagement facilitated by the show's online presence provides the opportunity for a profoundly different experience or relationship with the show than was available in previous eras. Viewers can feel that they have a relationship with the show and its cast, as well as with fellow viewers (extending the brand beyond ritual in-home viewing to its new role as facilitator of external relationships). The show also capitalizes on the types of experiences, information, and feelings that exist around women's magazines, gossip tabloids, self-help books, online discussion forums, and "mommy blogs," and connect them to the *Today* show. And, given that many women in the workforce have online access to the site at the office, *Today* has effectively expanded its reach beyond the constraints of the broadcast medium and daypart.

But as was argued earlier, crafting a different relationship with audiences is only one facet of the post-network reconfiguration involving talk shows. In late 2007, when NBC Universal acquired Oxygen Media for almost $1 billion, CEO Jeff Zucker proclaimed the network's intentions: "We want to create a virtual women's network where we go to market selling young women and affluent women in a way that virtually no one else can."[33] As Zucker correctly notes, networks are ultimately in the business of selling audiences, not producing entertainment for its own sake. It is a reminder that content in the post-network model becomes less important than, even secondary to, the search for audiences, even if that means gathering them together across disparate locations. In the past, the value of the "network" was as a brand name around which viewers would gather. A notable example, for instance, was the assumption of quality that would attract viewers to NBC's "Must See TV" on Thursday nights. The "virtual network" of broadcast, cable, and online media that NBCU has now created, by contrast, is a network crafted for advertisers. The network doesn't wait for valuable audiences to come to it but, rather, assembles audiences that advertisers covet *across* media. Beyond the purchase of Oxygen, NBC, for instance, built upon its popular *Today* show franchise to create a network of media properties that specialize in female-oriented content and programming. Other properties include Bravo Media, with its popular fashion-design show *Project Runway*, and iVillage, a web company with a robust community of women users. The organizational unit Women@NBCU handles this network of properties and serves as a consolidated means for selling female audiences to advertisers. The company, therefore, has the ability to target an advertiser's desired audience in "all the particular attributes and lifestyle and mindsets and brand objectives, and then come back here and … do a ton of research on our content across everyplace, and … actually put together a sort of virtual network just for them," says one NBC marketing executive. "When you put all of our properties together," she continues (perhaps overstating the case), "we reach almost every person in this country in a month, and so we reach every psychographic category."[34] In the network's vision, it is only through this "cobbling" approach of different media properties that it will be able to put together a mass audience in ways that approach what it was capable of in the past.

As for the effect on production, there are certainly visible content-sharing, distribution, and cross-promotion activities between these sites and locations. Experts from iVillage appear on air on *Today*, and the show's anchors often direct viewers to stories they can find on iVillage. Furthermore, the show's website includes articles that link directly to iVillage as well. Similarly, content from Oxygen has found its way into taxis and onto college campuses thanks to NBCU's powerful distribution reach.[35] Nevertheless, the changes that have occurred in the production of the *Today* show itself have been minimal. If one were to have watched the program ten years earlier, one would pretty much see the same essential structure and similar forms of content. The talk show's relationship to industrial reconfiguration, then, has less to do with altered production than in the institutional organization, distribution, financing, and cross-platform appeal that such arrangements allow. Furthermore, post-network expansion is the one area in which the company can demonstrate robust growth and additional revenue streams to Wall Street (an important factor in all publicly traded corporations). In the spring and early summer of 2007, NBC could boast a 150 percent increase in *Today* show webpage views from the previous year, with unique visitors up 200 percent.[36] NBC's digital media unit achieved revenues of $1 billion a year earlier than Wall Street forecasted. And traffic to iVillage increased by 2 million visitors in the two years since it was acquired by NBC Universal. Indeed, Zucker contends (as one industry publication summarized him) that "the era of growth in network TV … is over. Growing his business, then, means investing in cable, digital video, mobile—anything other than network TV."[37] Furthermore, "the network today makes up less than 5 percent of NBC Universal's earnings," while its cable division produces 50 percent of the entertainment division's revenue.[38] In sum, the digital strategy built upon the *Today* show's base audience has led the network to an expansion of properties that enable a different relationship to and different set of experiences for its audience, all the while utilizing that altered relationship as an intensified means of knowing and then selling its audience to advertisers.

Figure 1.2 The iVillage.com website offers numerous avenues for audience interaction, including across platforms, with the brand, and with other viewers.

Source: http://www.ivillage.com/ (accessed October 15, 2008).

While this discussion has focused largely on the *Today* show, the other networks and their flagship talk shows are undergoing similar changes. As noted previously, *Good Morning America*—the second highest ranked talk show in the morning daypart—expanded its programming to a cable- and digital-only third hour in the fall of 2007 as *Good Morning America Now*. The program is attempting to utilize its expanded digital space for new and extended segments not making it to air. It also touts its viewer-interaction features, such as airing user-produced promotional segments and "talk back" areas.[39] Similar to the Internet and cable expansions of the mid-1990s, digital efforts today can often be accompanied by rhetoric of viewer empowerment.[40] "I can be standing in the aisle at the supermarket," exclaims the show's executive producer, "and I can call up that segment [of Emeril Lagasse cooking] and voila, I know what products to buy .... Viewers are hungry for information they can take to the store. We will make that even easier for them."[41] Irrespective of whether viewers will utilize new media in this fashion, such rhetoric demonstrates the intentions of networks to sell advertisers on the potentialities of traditional media content distributed across new platforms. In short, *GMA Now* is an attempt by another television network to build on a "very, very robust and revered brand" and extend it further into viewers' lives by creating new relationships in new formats.[42]

Late-Night Snacks

Talk shows in the late-night daypart share both similarities with and differences from the morning shows and their role in the networks' digital strategies. Like morning shows, late-night talk is a programming form that is comprised largely of distinct segments, and thus has the potential to be broken into smaller video packages that the industry variously labels "snippits," "snackable content," and "video snacks" to refer to the ease with which they can be consumed in a variety of digital media platforms. Also similar to morning shows, late-night's websites have an archival quality, offering visitors the opportunity to watch entire episodes online and choose from an array of comedy and talk segments, as well as to engage in other features that might create a sense of community, such as signing up for podcasts, reading blogs and examining other "behind the scenes" activities, purchasing gear, and so forth.

One of the primary differences between late-night and morning talk, of course, is that late-night is first and foremost designed to entertain with humor, as opposed to the service orientation long a component of daytime talk. Late-night therefore maintains different forms of social currency and community-enabling features than morning talk, as well as distinctions in content. How such content will circulate outside the confines of the traditional broadcast venue, then, sometimes rests on this fundamental difference. Perhaps a viewer might want to see Al Roker on his cell phone, but David Letterman's "Top Ten List" is probably a more appealing talk segment for digital download. Furthermore, whereas the networks have attempted to exploit the morning talk shows by controlling distribution through ownership or proprietary sites, late-night content has been used much more promiscuously.

YouTube, in particular, has become a central location for the distribution and sharing of humorous talk programming. Jimmy Kimmel, host of ABC's *Jimmy Kimmel Live!*, posted a satirical video of his supposed sexual relationship with actor Ben Affleck on the site, and, according to one press account, it has been viewed over 8 million times.[43] NBC regularly posts clips of *Late Night with Conan O'Brien* and the *Tonight Show with Jay Leno* on Hulu.com (the video distribution site it owns jointly with News Corp.), as well as on YouTube. Leno even does an on-air comedy bit in which he offers funny and weird videos he has found on YouTube, thereby creating a circular effect of fans flocking to YouTube to watch both those clips and Leno's segments promoting those clips. *The Late Show with David Letterman* maintains an even larger presence on YouTube. When the show received a bump in its ratings a few months after clips began appearing there, CBS sent out a press release claiming that it had attracted 200,000 new viewers as a result.[44] After each show, the staff selects a single segment to post. But here, too, digital distribution doesn't seemingly affect content production. Letterman's writers contend that they're "really just looking at making our show as good as possible and then after the fact it's, 'All right, what do we want to put onto the internet.'" One writer notes, "We're always trying to find that water-cooler moment."[45]

But when digital media content is so easily reproduced and circulated, this statement begs the question of where or what *is* the water cooler in the digital age. The viral distribution of comedy talk segments suggests that it is perhaps in the interactive act of sharing—and the desire to do so—through which communal moments are now being created, whether through email, Facebook, or even a central site such as YouTube, rather than a physical location. Hearing someone repeat the joke heard on Conan is not the same as hearing it or seeing it yourself, as now possible. Humorous talk show content, then, becomes a communal commodity that viewers want to share (for laughing by oneself is rarely as fun). And this applies to talk interviews as well as comedic monologues and "Top Ten Lists." Letterman's interview with British comedian Sasha Baron Cohen (appearing in character as Borat) is one of the most widely viewed late-night clips on YouTube.

The type of humor found on talk shows is also generally an inoffensive form of amusement readily able to transcend the domestic space. It therefore is being widely used by the networks in their place-based media outlets. What Anna McCarthy described several years ago as "ambient television" has now, in the digital era, become an organized business practice in the distribution of video content[46] ABC, CBS, and NBC have all created out-of-home viewing networks

Figure 1.3 "Dave TV" allows viewers to watch almost all of the *Late Night with David Letterman Show* highlights in segmented ("video snack") form.

Source: http://lateshow.cbs.com/latenight/lateshow/dave_tv/highlights/index/php/bigshowhighlight.phtml (accessed October 15, 2008).

to sell advertising for the numerous screens available for viewing in locations other than the home. NBC, for instance, formed NBC Everywhere, which manages an array of sub-units such as NBC at the Gym, NBC on Campus, NBC in Taxi, continuing through outlets such as ballparks, supermarkets, schools, and hospitals.[47] CBS has a similar organizational arm called CBS Outernet that is carried in auto dealerships and service centers, grocery stores, and GameStop stores.[48] Here too is another enormous growth area for the networks and one that provides an opportunity to reassert the cultural centrality of television.[49] Late-night talk shows produce short and appealing video snacks that can be aired in a wide variety of viewing contexts such as these. Furthermore, by airing humorous talk segments from Craig Ferguson or Conan O'Brien, the networks can potentially grow their brand by attracting new eyeballs that are unfamiliar with these shows and their brand of comedy.

In sum, the value of the late-night talk show in the digital era may be in using its segmented form and appealing content for widespread distribution, thereby building up these franchises and the brand. While we may not be able to account fully for the 40 percent rise in late-night viewership in the last decade, especially given the newfound competition from cable competitors such as Jon Stewart and Stephen Colbert, the easy access to such programming through new and alternative video distribution outlets may be one of the contributing factors for the ratings increase. If viewers enjoy late-night talk and humor but the late-night hours prevent regular broadcast viewing, then alternative distribution allows for the extension of these brands into viewers' busy lives. Another value may be buzz. As the networks continue to bolster their infrastructure to support viewer and advertiser migration to new platforms, the widely recognized faces of Letterman and Leno and their signature comedy acts become a selling point to advertisers and viewers alike for these growth areas, from mobile phones to place-based media. Finally, the adaptability and appeal of humorous talk shows may allow for new means of community among viewers, as audiences have increased and enhanced opportunities to use this form of television content as social currency. The water cooler may be transformed by social networking technologies, even if the new ways of sharing what someone saw on television last night (or on the computer in the workplace) now occurs through the computer or cell phone in the workplace.

Conclusion

With morning talk shows expanding their offerings and late-night talk shows increasing in popularity, it is easy to see how the programming form is moving in the opposite direction to much of the rest of network fare. It is no surprise, then, that network executives have recognized this popularity and profitability and attempted to exploit it in their industrial reconfigurations. The discussions here have highlighted network expansion beyond broadcasting, as the networks are transforming themselves into media companies that extend across television (broadcasting and cable) into online, mobile, and place-based media. This expansion is institutional (acquiring properties) as well as programmatic (using content in a concerted fashion across platforms the networks don't control). As I have illustrated, late-night talk shows can be used to hail audiences in dispersed locations because of the familiar, appealing, and pleasurable nature of the content, thereby extending the brand and creating buzz for engagement with it. The networks have learned, too, to use morning talk as a vehicle to lead viewers into extended engagement with subject matter that they have always excelled in providing. By acquiring properties that also cater to the morning talk show demographic, the networks have demonstrated their ability to create, foster, and extend a new relationship with viewers. While morning talk "shows" have always been focused on creating a compelling and appealing performance to be experienced in real time, the "show" aspect of the programming may be becoming less important than its continued ability to draw viewers with similar interests, lifestyles, and behaviors and engage them for extended periods across media platforms. In both instances, the relationship with audiences has intensified, often centering on engagement with the brand. In short, simply supplying good content is no longer enough for the networks if they want to survive in the new media landscape.

The networks have also realized the need to employ new metrics in an attempt to profit from this altered relationship. New technologies allow for more intensive audiences measurement as viewers move from place to place, such as within the *Today* show website itself, but then on to iVillage, blogs, discussion boards, and so on. As the head of ESPN's digital and mobile media division noted about the industry as a whole, "You start to learn more and more about your fans as they migrate from platform to platform. What we're doing now is customer relationship management."[50] Whether this customer information will allow the networks to transform what are now digital pennies into digital dollars in the future is yet to be seen. But what is important to remember is that the networks no longer need to dominate the audience's attention through conventional television viewing in quite the same way as they once did. If they can facilitate movement that follows the content to other screens, perhaps even encourage it, then new opportunities exist for profit.

Talk shows, it seems, are a perfect programming type for the new digital universe. Beyond the qualities that are inherent to the form—such as segmented content that can be easily accessed through numerous technologies—talk and laughter are universal human behaviors that invite participation and engagement. The talk show increasingly becomes a programming form that can be "opened up" for broader viewer engagement as people are no longer forced simply to sit and watch such actions without actually participating in them themselves in social settings. While the talk show still fulfills the older network model of being an entertainment show that will regularly attract mass audiences, it is simultaneously a programming form that can be used to enhance viewer engagement with the network brand outside the confines of broadcasting. As we have seen, talk shows themselves are not being radically transformed by post-network reconfigurations. They are, however, participants in the radical transformation of audience engagement with television content in new and different ways.

Notes

The author would like to thank Amy Lutz-Sexton for her invaluable research assistance with this project.

1 Amanda D. Lotz, *The Television Will Be Revolutionized* (New York: New York University Press, 2007), 13.
2 Peter Johnson, "New Anchors Away!," *USA Today*, 29 August 2007, 1D.
3 Larry Dobrow, "Night Time the Right Time in Snaring Young Eyeballs," *Advertising Age*, 9 May 2005, S26.
4 Matea Gold, "Wake-up Call to A.M. News: Moms Tuning Out," *Los Angeles Times*, 9 February 2007, A1.
5 Ibid.; Michele Greppi, "Nets Milk Morning Show Cash Cows," *Television Week*, 3 September 2007, 60.
6 Paige Albniak, "Fourth Hour of *Today Show*?," *Broadcasting & Cable*, 5 February 2007. Retrieved from http://www.broadcastingcable.com/article/CA6413223.html.
7 Quoted in Bernard M. Timberg, *Television Talk: A History of the TV Talk Show* (Austin: University of Texas Press, 2002), 37.
8 As one CBS executive noted, "Seventy percent of our revenue comes from women 25–54." Quoted in Gold, "Wake-up Call", A1.
9 Ibid.
10 Bill Carter, "Gap Widens Between Morning TV Competitors," *New York Times*, 7 July 2008, E1.
11 Meg James, "For CBS the Web, Not Imus, is the News," *Los Angeles Times*, 16 April 2007, C1.
12 Bill Carter, "NBC Will Offer its Shows Free for Download," *New York Times*, 20 September 2007, A1.
13 Todd Spangler, "Can Comcast Deliver TV 2.0?," *Multichannel News*, 18 August 2008. Retrieved from http://www.multichannel.com/article/CA6587969.html?q=comcast+fancast.
14 James, "For CBS the Web," C1.
15 Shahnaz Mahmud, "ABC is Latest Net to Seek Web Alliances," *Adweek*, 10 September 2007.
16 CBS's actions in this regard have no bearing on its parent company's lawsuit against YouTube for past copyright infringement. Brian Stelter, "Some Media Companies Choose to Profit from Pirated YouTube Clips," *New York Times*, 16 August 2008, C1.
17 Karl Taro Greenfeld, "Zuckervision," *Condé' Nast Portfolio*, September 2008. Retrieved from http://www.portfolio.com/executives/features/2008/08/13/Profile-of-NBC-Universals-Zucker.
18 Daisy Whitney, "Nets Tinker with Social Networking," *Television Week*, 15 January 2007, 60.
19 Beth Comstock, "The 3 C's of Success in the New Digital Age," *Television Week*, 24 April 2006, 9.
20 Michael Learmonth, "Ladies First at NBCU," *Daily Variety*, 7 March 2006, 1.
21 Ibid.; emphasis added.
22 Steven Zeitchik, "Digging into Digital," *Daily Variety*, 31 May 2007, 5.

23 Abbey Klaassen and Andrew Hampp, "CBS, Spike Kick Video up a Notch," *Advertising Age*, 2 June 2008, 8.
24 Betsy Streisand and Richard J. Newman, "The New Media Elites," *U.S. News & World Report*, 14 November 2005, 54.
25 Zeitchik, "Digging into Digital," 5.
26 Ibid.
27 Klaassen and Hampp, "CBS, Spike Kick Video," 8.
28 Andrew Hampp, "Lauren Zalaznick; President of Women's Lifestyle Networks, NBC Universal," *Advertising Age*, 2 June 2008, S8.
29 For the argument that reality programming is, in many ways, a reformulation of traditional talk shows, see Jeffrey P. Jones, "Beyond Genre: Cable's Impact on the Talk Show." In Gary R. Edgerton and Brian G. Rose (eds.), *Thinking Outside the Box: A Contemporary Television Genre Reader* (Lexington: University Press of Kentucky, 2005), 156–75.
30 Ibid.
31 Comstock, "The 3 C's of success," 9.
32 Ibid.
33 Frank Ahrens, "Targeting Female Viewers, NBC to Buy Oxygen Media," *Washington Post*, 10 October 2007, D1.
34 NBC's Senior VP for Marketing, Debbie Reichig, quoted in Jon Lafayette, "Cobbling Together a Target Ad Audience, NBC Packages Programming Across Media," *Television Week*, 5 May 2008, 1.
35 Brian Stelter, "NBCUniversal Brings Oxygen into the Fold," *New York Times*, 16 June 2008, C6.
36 Greppi, "Nets Milk Morning Show," 60.
37 Greenfeld, "Zuckervision."
38 Ibid. Ahrens, "Targeting Female Viewers," D1.
39 Michele Greppi, "'GMA' Greets User Videos," *Television Week*, 18 February 2008, 6.
40 See Jeffrey P. Jones, *Entertaining Politics: New Political Television and Civic Culture* (Lanham, MD: Rowman & Littlefield, 2005), 40–52.
41 Jessica Stedman Guff, quoted in Michele Greppi, "'GMA' Taps Cuomo to Lead New Web Hour," *Television Week*, 3 September 2007, 1.
42 Ibid.
43 Edward Wyatt, "Late-Night TV Satires Become Online Hits," *New York Times*, 27 February 2008, E1.
44 Darel Jevens, "Dave Goes Digital by Accident," *Chicago Sun Times*, 4 March 2007, D1.
45 Ibid.
46 Anna McCarthy, *Ambient Television: Visual Culture and Public Space* (Durham, NC: Duke University Press, 2001).
47 Jon Lafayette, "NBCU Adds Gyms, Colleges to Lineup," *Television Week*, 21 January 2008, 29.
48 Brian Steinberg, "Big 3 TV Nets Follow Consumers out of the Living Room," *Advertising Age*, 11 August 2008, 4.
49 Wayne Friedman, "Tuning in to 'Promo Entertainment,'" *Television Week*, 19 May 2008, 40.
50 Louise Story, "Yes, the Screen is Tiny, but the Plans are Big," *New York Times*, 17 June 2007, C1.

Chapter 2

Like Sands Through the Hourglass

The Changing Fortunes of the Daytime Television Soap Opera

Elana Levine

The daytime television soap opera has long been a staple of American broadcast network programming, making a significant impact both economically and culturally. In the soaps' network-era heyday of the 1970s and early 1980s, the daytime schedule reportedly fed CBS and NBC 75 percent of their profits, and in 1981, just one of ABC's hit soaps, *General Hospital*, generated one-quarter of the network's yearly revenues.[1] Fifteen such programs aired daily during the 1981–2 season, filling the late morning and afternoon hours of the Big Three broadcast networks; a single soap could draw as many as 10 million households.[2] During this same period, daytime soaps became something of a national pastime: news magazines featured the programs on their covers, ancillary products from board games to T-shirts filled retail shelves, and hordes of college students designed class schedules and social gatherings around the daytime TV line-up. From the radio era of 15-minute serials into at least the early years of television's multi-channel transition period, the daytime soap was a vital player in the broadcast business model, serving as a valued vehicle for delivering female audiences to advertisers. But the soaps were also central cultural reference points for the multiple generations of audiences that faithfully followed characters and their story arcs over years and years of episodes.

As of the multi-channel transition era, however, the soaps began to experience a decline in popularity, profitability, and cultural centrality. Nielsen Media Research ratings have fallen from a high of around 12 for the most popular soaps in the early 1980s to a present-day high of around 4.2 (a drop from about 10 million to fewer than 5 million households), with most soaps rating between 2 and 3 on the Nielsen scale (2.2 to 3.3 million households).[3] Data on 18- to 49-year-old female viewers, the demographic advertisers are most eager to reach via the soaps, show declines in viewership as well. In the 1983–4 season, ABC was the network with the highest rating in this category, a 6.2 (about 5.2 million).[4] By 1990, ABC's ratings for 18 to 49-year-old women had decreased to 3.9 (about 3.6 million households).[5] Then, between 1994 and 2000, the

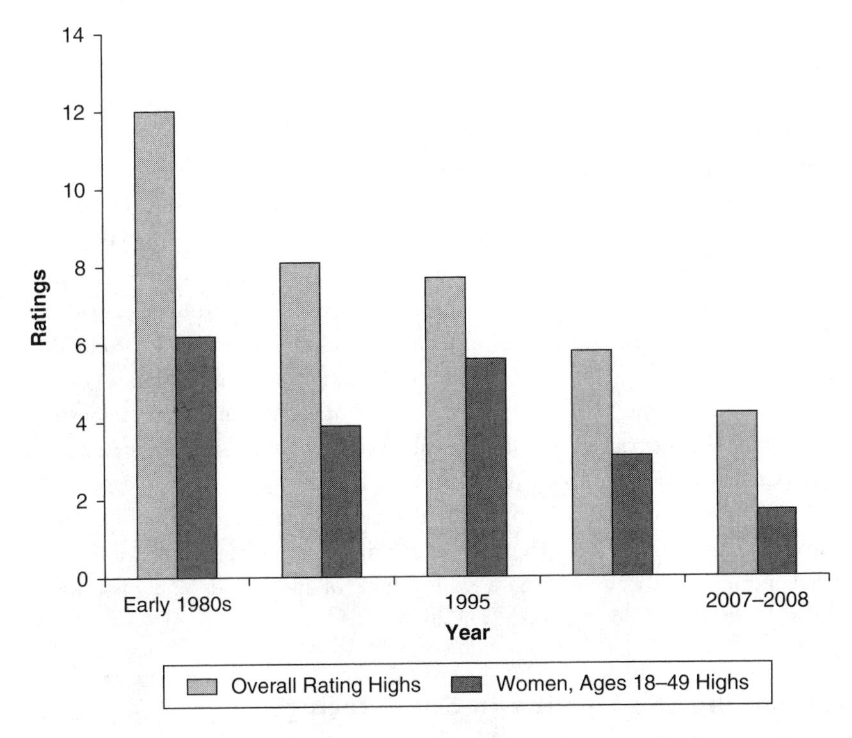

Figure 2.1 This chart illustrates the declining soap opera ratings among all audiences and among women aged 18 to 34 since the early 1980s.

numbers of this demographic watching all soaps declined by 27 percent,[6] and the numbers have continued to drop. Since the turn of the twenty-first century alone, NBC has faced a 35 percent loss in female viewers in this age group for *Days of Our Lives* and, in April 2007, ratings for this group for all soaps hit 1.5 (approximately 1.7 million viewers).[7] As of this writing, eight soaps remain on air, with several facing persistent rumors of cancellation.[8] While advertisers admit that daytime soaps still offer relatively cheap access to women viewers, and that no other media outlet can as yet deliver the same numbers for the same cost, advertisers, the broadcast networks, and soap producers all recognize the problematic position of this once-indomitable program form.[9]

The changing fortunes of the daytime television soap opera since the mid-1980s raise a number of questions about the past, present, and future of the genre. While I am not able to predict the soaps' future, nor interested in so doing, I can offer a detailed consideration of the various factors influencing their industrial and cultural position over the past 25 years, as well as an analysis of how soap producers, the broadcast networks, and audiences have responded to these shifts. Overall, this chapter examines why there has been such a significant drop in the soaps' popularity and profitability, how these programs and their viewers have responded to such changes, and how we might best understand the changes that are sure to continue. There are multiple reasons for the soaps' changing fortunes since the mid-1980s, and explaining these complex developments in terms of one catalyzing force or event—increases in working women or the disruption to daytime programming from the O. J. Simpson murder trial—does a disservice to our understanding of the workings of the US television industry and the practices of its viewers. Across such developments, network decision-makers and soaps' creative staffs have struggled to revitalize and sustain the soap genre—making changes in storytelling, in casting, in production models, and in practices of transmedia distribution and promotion—but the future security of this program form remains tenuous. While the precariousness of the soaps' industrial positioning has allowed for some experimentation and innovation in production and distribution, such changes have not necessarily been well received by long-standing audiences, nor have they worked to draw substantial numbers of new viewers. As a result, the future of a genre that has been immensely meaningful to a wide range of individuals and institutions throughout US broadcast history is in doubt.

Explaining the Decline in Soap Ratings

The steady, continuing decline in ratings for daytime soaps presents an alarming situation for the commercial media industries that allocate funding and make programming decisions based upon such data. Understanding the significance of this situation requires a consideration of the impact of these ratings drops upon advertisers, producers, networks, fan and popular press, and audiences, as well as the ways that such data affect these parties' attitudes toward soaps and their economic and cultural worth. As a result, in this section I examine the chief factors used to explain the decline in the ratings as offered by these various interests, but I do so with the recognition that each explanation is encumbered by the particular motives of those who offer it. My analysis thus focuses as much on those motives as it does on the explanations themselves.

The decline in soap ratings beginning with the multi-channel transition era is consistent with the ratings drops found across broadcast network television since that time. As Amanda D. Lotz has described, during television's network era, top-rated prime-time programs were watched by 40 to 50 percent of American households, whereas a top-rated broadcast of the early post-network era such as Fox's *American Idol* is watched by about 14 percent of US homes.[10] Since the initial fragmentation of viewing masses on account of the growth of cable and satellite services, all broadcast network TV has experienced ratings losses. With the onset of the post-network era, audience attention has been increasingly divided, now between "older" media such as television and "newer" media such as the Internet, even further eroding the ratings of daytime and prime-time broadcasting.

However, viewers turning to cable channels or to the Internet instead of the soaps is not the most prominent explanation offered by the soap industry or by journalists for the decline in ratings. That role is assigned to women leaving the home during the day to participate in the workforce, a trend most often noted as increasing markedly in the 1980s and which has become a *pro forma* acknowledgment in nearly all attempts to explain the soaps' decline. Indeed, from the mid-1960s to the mid-1980s the percentage of working women in the United States increased from 33 percent to 52 percent.[11] Yet the simplistic equation of more women working and fewer women watching soaps fails to address the real problem with these developments for advertisers. For the advertising industry (and hence for the networks that cater to them) more pertinent is the matter of the *kinds* of women they believe are too busy working during the day to watch soaps. In 2002, CBS research executive David Poltrack contended, "The No. 1 reason the demographics of daytime soap viewers are changing is the more mobile female population, with younger women in particular more likely to be part of the work force."[12] Thus, since the mid-1980s, advertisers, network executives, and the journalists who cover them have been largely concerned with the aging of soap viewers—one report claimed that their median age had risen by seven years between 1991 and 2002[13]—as well as their less desirable economic clout. As CBS researcher Arnold Becker explained as early as 1989, advertisers believe that the remaining soap viewers are "probably economically undesirable women, if you will—somewhat older, they are likely to be somewhat poorer, and maybe these are not the leading-edge women that they want to get."[14] Even earlier, in 1985, one executive admitted that advertisers' growing interest in prime time over daytime was because "We no longer want to reach the inner-city, less affluent viewer."[15] The racial as well as the economic dimension of such an admission points to the ways in which an off-hand assumption that "working women" have led to a decline in the soap audience ignores the fact that advertisers are particularly concerned with their ability to reach *a certain kind* of woman via the soaps: young, affluent, and white.

A second explanation for the shifts in the soaps' popularity and profitability also addresses the matter of the genre's presumably female audience. This explanation foregrounds the specific effect that soaps are believed to have upon that audience—an addictive power that is represented as explaining the lasting appeal of these programs. In such formulations, offered by soap producers, writers, and network executives as well as by the advertising industry and journalists, any disruption in the supply of this addictive substance can break the addiction, resulting in lost ratings points. This explanation for the soaps' decline in popularity appears throughout discussions of a number of forces competing for female viewers' attention, including the workplace, talk shows, and cable. But this discourse is exceptionally prominent from the mid-1990s on, when the disruption to the addiction becomes regularly attributed to the impact of daytime television broadcasts of the O. J. Simpson murder trial.

During Simpson's preliminary hearing in the summer of 1994, CBS, NBC, and ABC pre-empted their soaps for about a week. Then, for the 1995 trial, the Big Three at times interrupted daytime programming with trial news while Court TV and other cable channels broadcast the trial in full, drawing large audiences. Even before anyone had a clear idea of whether or not the trial would affect daytime drama ratings, soap producers and network executives invoked the language of addiction to express their concern that such a drop might occur. As *All My Children* executive producer Felicia Minei Behr explained, "Once you break the audience participation, it's difficult to get them to reinvest that time."[16] Journalist Larry Bonko even further articulated the soap fans' experience to that of addiction withdrawal:

> Viewers who don't give a hoot about Simpson's guilt or innocence but care deeply about what's happening with the much-married Erica Kane on *All My Children* will have their day ruined if the soap operas are preempted for the trial. From across the land, the networks and affiliates will hear the cry. "We want our soaps! We want our soaps!"[17]

Soap pre-emptions for the trial were minimal (and virtually non-existent for those shows that were broadcast during the court's lunch break) and the small dips in soap ratings that did occur were short-lived. Once the Simpson story settled down, ratings for daytime dramas returned to their "pre-O. J." levels.[18] And yet the "O. J. myth" has persisted when it comes to explaining the declines in soap viewing numbers. As *The Young and the Restless* writer Kay Alden described the trial's impact: "We never came back from that fully. There's a certain habituality to watching daytime, and when that habit is broken . . . ."[19] Similarly, *General Hospital* writer Michele Val Jean referred to the trial's habit-breaking impact when speaking about potential disruptions to the soaps because of the 2007 Writers' Guild strike: "Our audience watches because they've been watching for a long time. We lost 8 million viewers over the O. J. Simpson trial who never came back." *Soap Opera Digest*/*Soap Opera Weekly* editorial director Lynn Leahey reinforced the same idea, arguing that "once viewers lose the habit [of soap viewing], they often disappear for good."[20]

Regardless of the weak causal links between the Simpson trial and soap ratings declines, this explanation for the changes in soap popularity remains dominant because of its convenient fit with constructions of the soap audience as addicts whose main reason for watching is habit rather than conscious choice, loyalty, or pleasure. While there is no doubt that soap ratings have seen a rather steady decline since the mid-1980s, attributing that decline to the disruptive effect of an event such as the Simpson trial is a facile explanation that serves the needs of media industries operating within a patriarchal culture that refuses to accept that one might freely choose to watch such a de-legitimated, feminized cultural product.

Along with the vast range of viewing options available across cable and, now, the Internet, the impact of working women and events such as the Simpson trial are the most frequent explanations appearing in industry and popular discourse on the declining ratings for daytime soaps from the multi-channel transition era to the present. But other explanations circulate as well, some within the popular and soap press or in scholarly discourse, and some more exclusively within soap fan communities. One such explanation is that the kind of storytelling the soaps offer is no longer unique to daytime. This argument contends that, since the 1980s, prime-time programming has so thoroughly adopted the serialized narratives and ensemble cast structure of daytime drama that audiences can happily receive whatever pleasures such textual features might offer in the higher budget, higher cache, and less time-consuming world of weekly prime-time dramas and reality shows.

The history of this shift not only documents the move of soap-like storytelling into prime time but also displays some of the ways in which daytime soap opera figures as the other against which this "new" kind of high-caliber serialized narrative is defined. In 1974, Horace Newcomb argued that there was an "essential difference between daytime serials and evening episodic television," and that that difference was rooted in the soaps' use of time and their ability to tell stories over months and years, which led to "a different use of events and actions, different modes of characterizations, and finally to a different set of values than that found in most of television." Instead of seeing these differences as evidence of soaps' inferiority, however, Newcomb argued that these features demonstrated "what television art [could] be" in a way other program types could not.[21] The uniqueness and artfulness Newcomb identified in soaps depended in large part on their serialized structure, a narrative device mostly limited, when he wrote, to daytime. With the early 1980s efforts to target the working women audience, however, came the "soapoperafication" of prime time, a "sheer proliferation of prime time serial melodramas," including hit series such as *Dallas* and *Dynasty*, and a number of shorter-lived attempts as well.[22] In light of such developments, television scholars, as well as much of the television industry and the press that cover it, have seen this era as the beginning of the transfer of daytime's narrative features to prime time.

While this original proliferation of serial melodramas in prime time largely disappeared by the 1990s, serialization of various kinds has persisted, moving out from under the derogatory label of "melodrama" to the more rarefied one of "quality," a tendency presaged in the 1980s by such series as *Hill Street Blues* and *thirtysomething*.[23] By the mid-1990s, the abundance of "quality" serialized drama on television led critic Charles McGrath to herald the "triumph of the prime-time novel" in a *New York Times* appreciation piece, citing such series as *E.R.* and

N.Y.P.D. Blue, and juxtaposing them against the kinds of television he would not include in this "golden age": daytime talk shows, sitcoms, British dramas airing on PBS, and "prime-time soaps."[24] Television scholars picked up this sentiment as well, for example in Robert J. Thompson's anointment of the early 1980s to the mid-1990s as television's "second golden age" as a result of the prominence of quality prime-time drama (much of it serialized, with ensemble casts and characters with memories).[25] Similarly, in 2006 Jason Mittell declared that US prime-time television since the 1990s had reached a new level of narrative "complexity," visible primarily in its dramatic, serialized storytelling.[26] In such scholarly accounts, the daytime soap opera receives little mention or is otherwise rejected as being fundamentally different from the (typically) valued and praised prime-time programming under consideration, programming that may share some features with soaps but that has developed those features into a higher art form. When the features that made soap opera unique are transferred to prime time, such discourse suggests, soap opera ceases to have much, if any, artistic or cultural purpose; when the genre's storytelling style is no longer unique, it is constructed as even less significant than before.

In challenging some of these discourses, I am not trying to deny that prime- time programming has adopted the soaps' once unique trait of serialization, as well as some of their other features, such as melodramatic situations and ensemble casts. As with the other potential explanations for a declining cultural interest in soaps I have thus far discussed, however, I am urging that we take into consideration the particular interests at work in asserting such a shift. A reliance on this explanation, that soap-like storytelling has shifted to prime time and has thereby obviated the cultural place of soaps, whether true or not, must be recognized as being influenced by an implicit progress narrative that positions daytime soap opera as a more primitive form that has been culturally displaced by a higher quality, more complex version in prime time. In this respect such an explanation serves the interests of those who seek to validate and elevate certain kinds of prime-time television, whether they are television creators, journalist/critics, or scholars.

The one party whose explanations for the decline in the soaps' ratings and cultural centrality since the mid-1980s I have not yet considered is the audience, the viewers whose understanding of the soaps as texts is arguably as great as, if not greater than, that of those parties more typically given the authority to speak on this question. Fan discourse about what has happened to the daytime soap opera is the only site for explanations that grapple specifically with the *content* of soaps, with what stories they tell, through what characters, at what pace, and in what way. The soap press and blogosphere also address such matters, although the major print publications are rather reserved in their criticism of the genre as a whole. In any case, whether professional soap critics or not, those who speak to such questions actually watch daytime soap operas and offer explanations for their decline that relate directly to the ways the shows themselves have changed.

While viewers have specific complaints about particular soaps, the fan stance across the board is that, since at least the mid-1990s, daytime soap operas have failed to deliver the kinds of stories and characters that audiences want to see.[27] Consider this comment, written in response to a blog post addressing what has gone wrong with soaps in recent years:

> Back in the 70's & 80's I watched consistently because the characters and the stories offered me a sense of hope and perspective. No matter what was happening in my depressing childhood drama, the trials of Julie Williams, Kimberly Brady, Rachel Cory, or Eden Capwell were far greater. We didn't share the exact same problems, but their determination, perseverance, and resilience provided great great lessons. [*sic*] They never gave up, they never played victim, they just dealt with their problems day by day.
>
> In college I discovered perhaps the last golden age of daytime—ABC soaps in the first half of the 90's. Erica's addictions, Viki's personalities, Brenda's obsession with a mobster, all made fascinating, compelling, and inspiring entertainment.
>
> Now I only occasionally watch the remaining eight because they all lack these elements. There are no characters that I find compelling, no problems that I find realistic, no emotions I can relate to. The women that were once complicated and strong have been relegated to ... supporting the non-acting models, behaving like 20-year-olds, crazy mothers, or absent all together.[28]

This (former) fan here points specifically to the soaps' representations of women as that which has changed enough to drive her away from the genre. Other fans point to other textual characteristics: veteran actors fired or ignored, poor pacing (e.g., too quickly rushing a new couple together), or stories that ignore character histories. As one viewer wrote:

> I remember when Rachel and Mac's will they/won't they get back to together story on AW keep [*sic*] me intrigued at an early age. The Four Musketeers were teens being supported by familar [*sic*] adults on GL. There was conflict with emotions attached, and characters' story derived form [*sic*] their actions to situations. Now writers plan mureder [*sic*] mysteries with out [*sic*] deciding who did the deed. They change characters' personalities to suit a new story, and characters never grow and change form [*sic*] their past mistakes. Soaps are no longer a dramatic, romantic, fantasy inspiring entertainment escapsim [*sic*] for their viewers. They are quick, poorly produced sound bites that leave the viewers feeling empty. Just write good story, respect history and character, and most importantly respect the audience.[29]

Audience disappointment with the direction soaps have taken is palpable not only in these comments, but also everywhere fans—and former fans—congregate to discuss them. Much of this congregation now takes place online, where relative anonymity and groupthink can amplify the negativity and disgruntlement soap viewers express.

Still, the perspectives fans offer on the changes they have witnessed since the mid-1980s and especially since the mid-1990s are occasionally echoed by those who work in the industry, suggesting that these are not merely gripes of viewers but observations shared by many with a close understanding of the genre. Thus, soap actor Cady McClain has complained about the representations of women in recent years:

> I cannot understand why in a medium that caters to women, intelligent female characters have to be set up to be stupid and emotionally impulsive … "punished" for actions they would never have taken without a gun to their head … and left begging and pathetic . . . . I have to wonder what the hell is going on these days. Are strong, smart women so threatening that they have to be vilified?[30]

Similarly, soap actor Victoria Rowell has criticized daytime's racial representations, citing a lack of diversity in a genre that draws a large African-American audience.[31] Alongside such critiques of soap representations are the more general acknowledgments by those in the industry about what has gone awry in soap storytelling. Thus, long-time *One Life to Live* actor Robin Strasser argues that soaps are lacking "great story" and Paul Rauch, an executive producer of a number of different soaps from the 1970s through the early 2000s, points out that soaps have become "amorphous … diffuse … unfocused" as network executives have gained more power over creative decisions, increasingly asserting the networks' will over that of the shows' "chief creators" as the networks scramble to stem the flow of lost ratings.[32] Online fans, soap journalists, or those who work in production largely agree that the content of soaps has changed since at least the mid-1990s, driving continuing viewers away and deflecting any potential interest from new viewers as well. This explanation is awarded little to no credence in scholarly and journalistic discussions of the declines in soap popularity since the mid-1980s, in large part because the insights of viewers and the labor of workers receive so little recognition beyond the worlds of the fan-targeted press and blogosphere.

Most of the explanations I have thus far considered seek social and cultural causes for the declines in soap ratings. But it is also worth considering how those ratings are acquired, and what impact they ultimately have on the soaps' ability to generate advertising revenue. The potential inadequacies of past and present audience measurement systems point to the challenge of settling upon any definitive explanations for the soaps' recent history. Indeed, since at least the early

1980s, audiences have watched daytime soaps in a number of different ways, some of which have never been included in the industry's audience measurement systems. For example, viewers have long watched these programs communally, gathering in workplace lunchrooms, college dormitories, and other such sites, but such "out of home" viewing traditionally has not been measured by Nielsen Media Research. Although the company's initiative begun in 2007 to measure the "extended home viewing" of Nielsen families' college student children while away at school starts to address this viewing practice, it is still a limited measure, and one that comes 25 years into soap ratings declines.[33]

The high rate of time shifting associated with viewing provides a similar limitation to the ratings data about soaps. The growing numbers of women in the workforce coincided with the introduction of the VCR, and both coincided with the initial drops in daytime soap ratings. VCR recording has long been included in Nielsen's live program ratings data, but the company never developed a way to measure VCR playback, and certainly not to measure whether or not viewers watched advertisements during playback.[34] Because of this, advertisers have long been wary of accepting ratings numbers that include VCR recording as an accurate count of viewers.[35] This has placed soaps in a difficult position for advertising sales, in that advertisers are less willing to "count" VCR recording than are the networks, and VCR recording composes a substantial portion of the soaps' ratings. As late as 2007, VCR recording made up between 4 and 12 percent of the live program rating for each daytime soap.[36] In the late 1980s, when VCRs had broader penetration and use (as they have now been replaced in some homes with DVRs), daytime soaps were the second most frequently videotaped type of program; soap industry pioneer Agnes Nixon even claimed that her creation *All My Children* was television's most-taped show.[37] But such claims did little to impress advertising buyers, then or now.

The new technologies of the post-network era have brought some of the inadequacies in Nielsen's handling of the VCR to light, in that DVRs have made all parties more cognizant of time-shifting practices and their impact on consumer exposure to commercial messages. Nielsen began measuring DVR playback at the very end of 2005, but these data did not enter into use in the buying and selling of commercial time until May 2007, when the company began to offer average ratings for the commercial minutes in each television program. Such data were available for programs watched live, but also at the levels of live plus same-day DVR playback, and DVR playback within one, two, three, and seven days.[38] After much negotiation, broadcasters, cable channels, and the advertising industry agreed to use average commercial minute ratings for the "live plus three days" playback measure as the standard for advertising buys.

As with VCRs, the high rate at which viewers time-shift daytime soaps via DVR makes DVR measurement especially significant for their ratings. As global media agency Carat reported after the first week of this new measurement, "For key daytime target, Women 25–54, average commercial minute ratings including three days of DVR playback (AKA 'C3') outperformed the live television rating for nearly every single daytime show."[39] Indeed, daytime is the daypart with the largest increase in ratings (10.1 percent) as a result of the new measure.[40] However, the average commercial minute measure and the three-day playback framework limit the extent of daytime's gains, further shaping the impact of audience measurement data upon conceptions of the soaps as a declining genre. To be included in the C3 measure, the Nielsen household watching soaps via DVR must play rather than fast-forward commercials, and must watch soaps within three days of recording. Nielsen data indicates that 85 percent of daytime soaps recorded via DVR are played back the same day,[41] but anecdotal discussion by soap viewers suggests that it is not uncommon to watch a week's worth of episodes in a weekend. Herein lies yet another disconnect between viewer practices and audience measurement systems. Advertiser interest in DVR playback declines the further it occurs past the live broadcast, in some cases because the advertising message is time-specific, as in promotions for a weekend sale, or in others because advertisers believe that viewers are unlikely to watch a show that airs five days a week on much of a delay. They assume that viewers would "fall hopelessly behind" and would not bother to watch at all.[42]

With the complications of a fallible audience measurement system on top of the range of social and cultural explanations for the soaps' diminishing centrality, all that is clear is that no one reason adequately explains the declines in ratings and in the seeming popularity of daytime soaps since the mid-1980s. Neither women in the workforce nor the O. J. Simpson trial, neither the adoption of serialized storytelling in prime time nor changes in soap plots and characters can wholly explain the diminishing interest in soaps in American culture at the end of the twentieth and the beginning of the twenty-first century. Taken together, however, such explanations begin to paint a fuller picture of the multiple factors at work. Although recent ratings data put a somewhat more positive spin on the soaps' position, the post-network era has been one in which the broadcast networks and soap producers have adopted a number of new tactics in a somewhat desperate attempt to sustain this once-vibrant program form.

Soaps' Effort at Survival in the Post-Network Era

Producers and network executives have not readily accepted the changing fortunes of the daytime soap opera. Instead, they have worked to attract viewers, increase ratings (or at least halt their continuing decline), and appease advertisers. They have approached these goals through a number of methods, including changes in content, in production practices, and in promotional tactics. The ultimate result of these strategies—in effect, their ability to save the soaps from extinction—is yet to be determined. But the industry's efforts along these lines can be seen as falling into two chief categories: spending less money on production and finding new streams of revenue. The hope is that these efforts will make soaps more cost efficient while attracting young, affluent viewers. While such efforts may accomplish some of the industry's economic goals, their reception among existing soap viewers and by many of the industry's creative workers has been less positive.

The soaps have worked to cut production costs in a number of ways in the post-network era. One such tactic has been the way the serials have handled the contracts of a number of their actors, particularly veteran performers whose long tenures have resulted in substantial salaries but whose age is believed to be a deterrent to the young viewers desired by advertisers. Throughout the 2000s, multiple soaps have fired some of these actors, reduced their minimum guarantee (the number of episodes per week in which they are guaranteed to appear on screen, or at least to be paid), or moved them from contract to recurring status (meaning that they are paid only a per-appearance fee with no guarantee).[43] These budget reductions have specific textual consequences. As critic Patrick Erwin argues, the recurring status of much of the *Guiding Light* cast as of 2008 keeps the writers from scripting long-term stories that involve multiple characters, as it is never clear whether recurring actors will be available to play out the story in the future or whether the budget will be able to pay for them.[44] Younger actors, new to daytime, are also being affected by such changes in casting practices, in that industry pay averages have been trending downward since the mid-1990s. According to talent manager Michael Bruno, in the 1980s and early 1990s, "As a new player in your first contract, you were getting up to at least $1,600 at two [episodes per week] and if they really wanted you, you could get up to $2,000 a show." More recently, however, newcomers may make $900 to $1,100 per episode, sometimes with just a one and a half episode per week guarantee.[45]

Cost-cutting measures have affected not only casting, but other aspects of soap production as well. As of 2008, only three serials (of the remaining eight shows on air) still employed breakdown writers, the scribes who take the head writer's long-term story projection and plot out how the story will "break" across months and weeks and days. Because, as former soap writer Tom Casiello argues, "It is much cheaper to not have breakdown writers, and everything is about slashing the budget now," head writers are taking on what have historically been two separate jobs, with less time available to devote to either one.[46]

The most dramatic example of the soaps' cost-cutting measures is the radical change to the *Guiding Light* production model that debuted on air on February 29, 2008. Facing some of the lowest ratings and oldest average viewers in daytime, *GL* producer Procter & Gamble sought simultaneously to reduce production costs and to appeal to younger viewers by helping the show "feel more real," more akin to serialized reality programs such as MTV's *The Hills* which have achieved youth loyalty in the post-network age.[47] To do so, the serial overhauled the production model that US daytime soaps have used since the 1950s. Instead of using two- and three-walled studio sets necessarily rotated within the constrained space of the program's New York City studio, the serial constructed 40 permanent sets in four-walled rooms with ceilings, as well as taking about 20 percent of all production on location in a nearby New Jersey town. This shift offered distinct savings in labor costs, in that the serial would no longer need an overnight crew to strike and assemble sets.[48] Nor would it need as extensive a lighting crew, particularly with the large amount of outdoor shooting and the accompanying switch to wireless, handheld, digital cameras that rapidly transmit footage to non-linear digital editing systems.[49] Freed of 300-lb. studio cameras and large, round-the-clock crews, the new *GL* streamlines personnel and budgets while maximizing space and, producers claim, extending the "realness" of the Springfield presented to viewers.[50] When ratings remained steady one month after episodes produced in the new model began to air, *GL* received a stay of execution, with CBS and Procter & Gamble announcing that the more cost-efficient serial would be renewed through 2009.[51] However, rumors of cost overruns and behind the scenes disgruntlement continue to plague the show, and online fan discourse has noted little improvement in the storytelling areas they see as most crucial to the genre's success.[52]

The soap industry has also focused on developing new methods for generating revenue. Chief among these are efforts at product placement and integration motivated by advertisers' desires to build deeper loyalties within those audiences that have remained with the shows. Lotz has explained that such strategies have become increasingly common since 2001 and that daytime soaps were among the first scripted series to feature attempts at "organic" integration of brands.[53] Indeed, the new, "realistic" feel for *Guiding Light* has been promoted as especially well suited to product placement deals, in that the appearance of real-world, name-brand products ostensibly seems more organic in the real-world environment created on screen. Increasingly, all of the soaps have been experimenting with such efforts. For example, in February 2008, the Campbell Soup Company sought to enhance its public image by sponsoring a campaign for women's awareness of heart disease, "communicating the benefit of their brands," and partnered with ABC Daytime to do so across its program line-up.[54] On *General Hospital*, Campbell's products were integrated throughout the month—orderly Cassius remarked that he needed to "drink his V8 Fusion" to keep up with a demanding nurse, and attorney Diane noted her friend Alexis's use of "Prego, the working woman's friend!" The brand and the heart health message were most closely integrated with the scripting of a heart attack for Epiphany Johnson, a middle-aged, overweight African-American nurse (a recurring character who is not part of any ongoing storyline). After surviving the attack, Epiphany grudgingly met with other women survivors of heart disease. As the women discussed their experiences, they also praised Campbell's heart-healthy choices and noted the company's sponsorship of their organization.

Although a successful integration for Campbell Soup and ABC, soap critics' irritated response to this placement illustrates the different foci of the industry and its audiences as the genre struggles for survival. While ABC Daytime found this a means of generating revenue, viewers identified it as highlighting the storytelling problems they saw as already plaguing the soap.[55] Epiphany's heart attack, particularly the way in which the story was told within a week's time and was so clearly a device meant to forward the sponsor's message rather than to explore character, seemed all too much in keeping with the show's narrative missteps in recent years. The fact that the heart attack victim was Epiphany, a character that fans have been eager to get to know better, but whom producers seem to have little interest in developing as more than a sassy voice of wisdom to motivate the actions of the young, white, conventionally attractive front-burner characters, exemplified soap viewers' ongoing complaints about representational politics and the lack of care with which soap narratives are told in the post-network age.

Alongside product integrations, the daytime soaps of the post-network era have also sought to generate new revenue and attract new viewers by working to maximize the shows' transmedia presence, exploiting media convergence to offer cross-platform distribution windows as well as developing new promotional outlets designed to drive viewers back to the TV screen. In 2000, Disney–ABC began the Soapnet cable channel as one such effort, originally replaying ABC's daytime dramas in the evenings and in weekend marathons, as well as rerunning some "classic" canceled soaps. By 2003, all three broadcast networks were experimenting with web-based content, initially through standard promotional sites and discussion forums. And, as of 2005, the networks began offering access to the soaps online in unprecedented ways. These strategies included selling merchandise branded with a soap's name or worn by soap characters and presenting daily quizzes about story content to reward regular viewing with "points" applicable to retail purchases. But the networks have also offered viewers new ways to consume the soap narratives themselves. For example, NBC established a special website in which viewers could discover additional "clues" to *Passions*' "Vendetta" mystery, CBS offered audio podcasts of *Guiding Light* and *As the World Turns* episodes, and ABC sponsored blogs authored by characters on each of its three soaps. As broadband capacity improved and streaming video became more viable by 2006, the networks also began to offer more video content online, as in video podcasts for ABC's *All My Children*, an online reality show, *InTurn*, in which contestants competed for a role on CBS's *As the World Turns*, and streaming daily episodes of CBS's regular line-up of four soaps.

All of these ventures seek to improve the soaps' financial standing by providing new outlets to sell advertising or ancillary materials and, ultimately, to drive audiences back to daytime viewing and thereby raise ratings and advertising revenue. As Josh Griffith, executive producer of *The Young and the Restless*, has said of the online magazine *Restlessstyle.com* (the real-world website that is the brainchild of fictional characters in *Y&R*'s Genoa City), "I think that wherever we can go to reach the largest number of people and get them excited about coming back to our show, you gotta do it."[56] CBS senior vice-president of daytime programs Barbara Bloom explains the range of multi-platform offerings similarly: "All of this is designed to keep the audience invested in the primary source, which is the daily show."[57] Yet the addition of ad-sponsored streaming episodes of the full CBS daytime line-up since 2007 suggests that such outlets may become revenue-generators of their own.

The Disney–ABC cable channel Soapnet now stands as an industry model for cross-platform distribution that generates revenue and has seen steady growth in subscribers and ratings since its 2000 debut.[58] In 2007, a thirteen-week prime-time series, *General Hospital: Night Shift*, pulled exceptionally high ratings for a niche cable series, reaching just under a million viewers and also achieving "respectable" numbers among the 18–49 demographic, not to mention surpassing the ratings for Soapnet's nightly repeats of *General Hospital* itself. The channel's success in 2007 was assisted by its strong showing with the new average commercial minute ratings measure, and resulted in a 40 percent increase in revenue in that year's upfront advertising market, as well as 24 new advertisers.[59] The fact that ABC owns its soaps as well as this cable outlet allows the company to profit from the content it produces in multiple ways. Disney–ABC daytime president Brian Frons presents it as a "holistic revenue-driven model" that focuses on getting as many people watching as possible across broadcast daytime, Soapnet, and international platforms. He argues that this is what makes the product appealing for advertisers—an efficient means of reliably reaching women consumers.[60] According to Frons, "We're able to aggregate those revenues ... and have a healthy bottom line."[61]

While soap fans have been increasingly disgruntled with Soapnet, upset at its inclusion of serialized prime-time dramas such as *The O.C.* and *One Tree Hill* as well as reruns of made-for-TV movies in place of repeats of classic daytime soaps, the channel has thus far allowed Disney–ABC to stay afloat in a market of otherwise depressed economics. The resentment of these strategies among at least a segment of soap viewers may one day harm the thus far "healthy bottom line" Frons identifies, but in a post-network context of uncertainty, in which advertisers strive to make sense of new ratings data and determine the most cost-effective ways to reach desirable consumers, the priority of the soap industry is not to deliver satisfying storytelling but rather to manage the "monetization of the cumulative ratings" it does get.[62] Thus, while the economics of a corporate structure such as Disney–ABC's are working, in the present, to sustain the soap genre, it is as yet unclear how that genre might change to suit its new corporate model.

There is no doubt that the daytime soap opera has been a vital fixture of American broadcasting since its early years, and that that kind of fixity must be due to the unique combination of the form's profitability to the media and advertising industries that have produced, sponsored, and aired it and its meaningfulness to the audiences that have so loyally consumed it. While the multichannel transition and post-network eras have brought with them a number of social and economic changes that have altered and may continue to alter the fortunes of the daytime television soap opera, no one factor will determine the genre's demise, just as no one factor can bear the responsibility of the ratings declines since the multi-channel transition. What this period of change continues to demonstrate, however, are the powerful community bonds that cohere around soaps, as can be seen in the impassioned criticisms of recent shifts in storytelling and production by invested viewers and industry workers. The sense of change surrounding the daytime soap opera is clearly felt by all of the relevant parties, and change can often bring with it feelings of anxiety, defensiveness, or panic. That these intense responses have come from the broadcast networks, soap writers and producers, soap performers, and soap fans alike over the multichannel transition and post-network eras testifies to the deep investments of both the television industry and the culture within which it operates in the ongoing, if imperiled, tale of the daytime television soap opera.

Notes

1 Bill Greeley, "Daytime Now 3-Web Scramble," *Variety*, 1 December 1971, 29; James P. Forkan, "Will Writing Team Find Happiness?" *Advertising Age*, 5 October 1981, 3; "She's Behind the Biggest Bubble in Showbiz: Soap Operas," *People*, 4 January 1982, 44.
2 Christopher Schemering, *The Soap Opera Encyclopedia* (New York: Ballantine Books, 1985), 313.
3 I calculated the number of households by multiplying the published ratings with the number of households represented by each ratings point at the different historical moments I mention and then rounding off those figures. The number of households represented by each ratings point has steadily increased over time, making ratings data in and of themselves an inaccurate measure of the changing size of the audience. Because the overall ratings measure households rather than individual viewers, it is difficult to be more specific in terms of actual numbers of viewers, especially without access to the proprietary reports Nielsen sells to its subscribers.
4 Peter Kerr, "Daytime TV's Ratings Drama," *New York Times*, 18 August 1984, 31.
5 Stephen Battaglio, "Viewers Losing Love for ABC's Afternoon," *Mediaweek*, 13 August 1990.
6 Lisa Leigh Parney and M. S. Mason, "Selling Soaps," *Christian Science Monitor*, 7 July 2000.
7 John Consoli, "Sun Setting on Key Daytime Demos," *Mediaweek*, 14 November 2005, www.mediaweek.com/mw/news/recent_display.jsp?vnu_content_id=1001478636. John Consoli, "Soaps on the Ropes," *Mediaweek*, 2 April 2007, http://www.mediaweek.com/mw/current/article_display.jsp?vnu_content_id=1003565868.
8 A ninth soap, *Passions*, left NBC for DirecTV in September 2007 and aired its last episode in August 2008.
9 Consoli, "Soaps on the Ropes."
10 Amanda D. Lotz, *The Television Will Be Revolutionized* (New York: New York University Press, 2007), 43.

11 Julie D'Acci, *Defining Women: Television and the Case of Cagney & Lacey* (Chapel Hill: University of North Carolina Press, 1994), 231, n. 13.
12 Pamela Paul, "Soap Operas Battle the Suds," *American Demographics*, January 2002.
13 Ibid.
14 Diane Haithman, "The Reselling of Daytime Television," *Los Angeles Times*, 31 August 1989.
15 Ronald Alsop, "Advertisers Go Beyond Soaps to Reach Daytime Audience," *Wall Street Journal*, 19 September 1985.
16 Connie Passalacqua, "Post-trial, Is There Life for the Soaps?" *Los Angeles Times*, 15 August 1995, 1.
17 Larry Bonko, "Soap Fans Need Not Dread Trial," *Virginian-Pilot*, 19 January 1995, E2.
18 Steve Weinstein, "Soap Operas Are Bubbling, Bouncing Back from O. J.," *Los Angeles Times*, 21 December 1995, 2.
19 David Finnigan, "Soap Selling Heats Up," *Brandweek*, 19–26 August 2002, 36–39, 40.
20 Lynn Smith, "A Cliffhanger for Soaps," *Los Angeles Times*, 12 November 2007, 11.
21 Horace Newcomb, *TV: The Most Popular Art* (New York: Anchor Books, 1974), 163.
22 D'Acci, *Defining Women*, 72. Jane Feuer, *Seeing Through the Eighties* (Durham, NC: Duke University Press, 1995), 114, 129–30, n. 3. Feuer is adding to Newcomb's conception of the soap opera here by noting not just the increased serialization but also the adoption of melodramatic storytelling in prime-time television of the 1980s.
23 Feuer argues that such "quality" series want to deny their link to the daytime soap opera, however: "Although one [the quality series] claims to be art, the other [the daytime soap] trash, they share in common a tendency that pervaded American television of the 1980s: serialization." Feuer, *Seeing Through the Eighties*, 111.
24 Charles McGrath, "The Triumph of the Prime-Time Novel," *New York Times* (Sunday Magazine), 22 October 1995, 53. McGrath also praises episodic series, particularly *Law & Order*, as attaining a level of novelistic quality, but those programs with some degree of serialization receive his greatest attention and approbation.
25 Robert J. Thompson, *Television's Second Golden Age* (Syracuse, NY: Syracuse University Press, 1996).
26 Jason Mittell, "Narrative Complexity in Contemporary American Television," *Velvet Light Trap*, 58 (Fall 2006), 29–40.
27 When I refer to soap audiences here and throughout the chapter, I am basing my claims on my ongoing survey of the online discourses of soap viewers, as seen in blogs and on websites, message boards and discussion forums, and podcasts. While the opinions of all soap viewers may not be represented in such forums, the extensive agreement across such sites on the "what is wrong with soaps" question, and the continuing nature of such discussions over many years, has convinced me that such views represent a significant and shared perspective.
28 Fabobug, comment posted 23 September 2007 to Marlena de Lacroix, "O. J. Simpson Didn't Do It! (Kill the Soaps, That Is)," *Savoring Soaps*, 19 September 2007, http://blogs.mediavillage.com/savoring_soaps/archives/2007/09/oj_didnt_do_it.html.
29 Lightkeeper, comment posted 17 September 2007 to "Daytime is Doomed, Part 3,013," *Snark Weighs In*, 16 September 2007, http://snarkweighsin.blog-city.com/daytime_is_doomed_part_3013.htm.
30 Quoted in transcript of *Soap Opera Weekly* interview with McClain, 3 January 2008, accessed at http://tadndixie.com/forums/index.php?showtopic=3813.
31 Nelson Branco, "Victoria Rowell Uncensored, Part II," *TV Guide Canada*, 18 December 2007, http://tvguide.sympatico.msn.ca/Victoria+Rowell+Uncensored+Part+Two/Soaps/Features/Articles/071217_victoria_rowell_NB.htm?isfa=1. While African-Americans make up 12.4 percent of the US population, they make up 20 percent of the daytime television audience. Rowell's former soap *The Young and the Restless* attracts 15 million African-American viewers, and all of the CBS daytime line-up receives high ratings among this group. Carolyn M. Brown, "Daytime's Other Drama," *Black Enterprise*, 1 December 2004, http://goliath.ecnext.com/coms2/gi_0199–3609642/Daytime-s-other-drama-despite.html.
32 Jennifer Lenhart, "How to Save Soaps," *Soap Opera Digest*, 25 September 2007, 39; Paul Rauch, personal communication, 25 October 2007.

33 For discussion of this adjustment in Nielsen's research, see Lotz, *The Television Will Be Revolutionized*, 204–5. Early reports of these extended ratings showed improvements for a number of daytime soaps. Rick Kissell, "College Campuses Boost Ratings," *Forbes.com*, 15 February 2007, http://www.forbes.com/digitalentertainment/2007/02/15/cx_rk_0215varietytv.html.
34 Larry Goldstein, "The Debate Over VCR Data—Baked into Nielsen Ratings or Not?" *Mediaaudit.com*, 12 August 2006, www.mediaaudit.com/downloads/mmi_debate_vcr.pdf.
35 Jon Lafayette, "Skeptical Buyers Await DVR Stats," *Television Week*, 21 November 2005.
36 Nielsen Media Research, "Network Daytime VCR Recording Contribution to Household Rating," Week of 9/17/07, chart linked from "Looking at Week 4; Daytime C3," *Radio Business Report/Television Business Report*, 29 October 2007, http://www.rbr.com/tv-cable/3042.html. The chart itself appears at http://www.rbr.com/epaper/images/102907-charts1.jpg.
37 Barbara J. Irwin, "An Oral History of a Piece of Americana: The Soap Opera Experience," PhD diss., SUNY Buffalo, 1990, 30.
38 Nielsen Media Research, "Nielsen Launches Commercial Minute Ratings in Standardized File," 31 May 2007, http://www.nielsenmedia.com/nc/portal/site/Public/menuitem.55dc65b4a7d5adff3f65936147a062a0/?x=8&show=%252FFilters%252FPress%252FNews%2BReleases%252FGeneral&vgnextoid=7f3a957da42e2110VgnVCM100000ac0a260aRCRD&from=01%252F01%252F2007%257C12%252F31%252F2007&y=8&selOneIndex=0.
39 Quoted in "Looking at Week 4; Daytime C3," *Radio Business Report/Television Business Report*, 29 October 2007, http://www.rbr.com/tv-cable/3042.html. The only daytime show for which this was not the case was game show *The Price is Right*.
40 "CPMs on the Rise for Network TV," *Media Buyer Planner*, 24 September 2007, http://www.mediabuyerplanner.com/2007/09/24/cpms-on-the-rise-for-network-tv/.
41 Paul J. Gough, "Nielsen on the DVR Watch," *Hollywood Reporter*, 16 February 2007, http://www.allbusiness.com/services/motion-pictures/4912255–1.html.
42 This logic has made some daily programs, such as syndicated talk shows, more desirable to advertisers, who believe the programs are more likely be watched live. Michael Learmonth and John Dempsey, "Cable Up in Upfront," *Daily Variety*, 2 July 2007.
43 Elaine G. Flores, "An Insider's Glimpse at Contract Negotiations," *Soap Opera Digest*, 7 March 2006, http://www.mbgla.com/press_detail.aspx?id=206.
44 Patrick Erwin, "Why *Guiding Light* is Self-Destructing," *Marlena De Lacroix*, 16 May 2008, http://marlenadelacroix.com/?p=91.
45 Deanna Barnert, "Hu$h Money," *Soap Opera Digest: Soap Opera Secrets*, 28 March 2005, http://www.mbgla.com/press_detail.aspx?id=187.
46 Tom Casiello, "My Nervous Breakdown," 18 April 2008, *Tommy C MySpace.com blog*, http://blog.myspace.com/index.cfm?fuseaction=blog.view&friendID=31817970&blogID=381333362.
47 Ed Martin, "*Guiding Light* Revolutionizes Daytime Production Models as it Enters the Digital Age," *JackMyers.com*, 14 February 2008, http://www.jackmyers.com/commentary/ed-martin-watercooler/15634227.html. Don Kaplan, "That was Then. This is Now!" *New York Post*, 16 October 2007, http://www.nypost.com/seven/10162007/tv/that_was_then_and_this_is_now.htm.
48 Even the production offices, bathrooms, and stairways of the *GL* studio do double-duty as sets in the new production model. Martin, "*Guiding Light* Revolutionizes Daytime . . .."
49 A *GL* location shoot in this new model requires a crew of nine or ten, compared with the 60 people necessary for a location shoot under the old model. Meg James, "Soaps Scrubbing Old Look," *Los Angeles Times*, 29 January 2008.
50 Brian Steinberg, "Soaps: They're Not Your Mother's Daytime Dramas Anymore," *Boston Globe*, 27 April 2008, http://www.boston.com/business/articles/2008/04/27/soaps/?page=1.
51 Don Kaplan, "*Guiding Light* Saved by *The Hills* Style," *New York Post*, 19 March 2008, http://www.nypost.com/seven/03192008/tv/guiding_light_saved_by_the_hills_style_102630.htm.
52 "Backstage Turmoil," *Soap Opera Weekly*, 3 June 2008, 10. Patrick Erwin, "Why *Guiding Light* is Self-Destructing," *Marlena De Lacroix*, 16 May 2008, http://marlenadelacroix.com/?p=91.
53 Lotz, *The Television Will Be Revolutionized*, 169–70.

54 John Consoli, "ABC Daytime, Campbell Ink Pact," *Mediaweek.com*, 28 January 2008.
55 Becca, "*General Hospital* Couple of Weeks in Review: As Usual, I Have Questions," *Serial Drama*, 10 February 2008, http://64.233.169.104/search?q=cache:MGgtJj_3UXcJ:serialdrama.typepad.com/serial_drama/2008/02/general-hospi-1.html+V8+fusion&hl=en&ct=clnk&cd=1&gl=us. See also Jacci Lewis, "*General Hospital*'s Heart Disease Awareness Effort Strikes an Awkward Note," *TV Envy*, 7 February 2008, http://television.gearlive.com/tvenvy/article/q107-general-hospitals-heart-disease-awareness-effort-strikes-an-awkward-no/.
56 Abby West, "*Y&R*'s Fictional Magazine Launches Actual Website," *EW.com*, 27 May 2008, http://www.ew.com/ew/article/0,,20201382,00.html.
57 Christopher Lisotta, "Soaps Shine as Lab for Ventures," *Television Week*, 17 July 2006.
58 Heidi Vogt, "Rising Suds Over at Soapnet," *Media Life Magazine*, 29 January 2003, http://www.medialifemagazine.com/news2003/jan03/jan27/3_wed/news5wednesday.html. Jon Lafayette, "Originals Clean up Soapnet's Summer," *TV Week*, 12 August 2007, http://www.tvweek.com/news/2007/08/originals_clean_up_soapnets_su.php.
59 Lafayette, "Originals Clean up Soapnet's Summer."
60 Anne Becker, "Disney–ABC Owns the Day," *Broadcasting & Cable*, 31 March 2008, http://www.broadcastingcable.com/index.asp?layout=article&articleid=CA6546060.
61 Joseph Adalian, "Kudocast Channels Daytime's Traumas," *Daily Variety*, 11 May 2007.
62 Ibid.

Chapter 3

The Benefits of Banality

Domestic Syndication in the Post-Network Era

Derek Kompare

> If you see syndication as a brand of national TV, it's an invisible brand that's been sold as a collection of programs. But what does syndication itself stand for? Somehow cable stands for something: It's automatically associated with upscale, targeted, efficient and added value. But if you say syndication, I'm not sure what people think.[1]
>
> Syndication is the kind of girlfriend your mother always wanted you to have—she's stable, not sexy.[2]

The first decade of the twenty-first century has been among the most eventful periods in the history of television. Yet most critical assessments attend toward the dynamic programming, business strategies, and technologies of the broadcast and cable networks, while syndication—the non-network, non-cable distribution of programs directly to local stations—is scarcely considered. It is not difficult to see why. Syndicated programs are almost nobody's favorite forms of television. Regarded in popular and critical imagination as the déclassé domain of aging reruns, mawkish talk shows, vapid games, and exploitative celebrity and court (and even celebrities-*in*-court) shows, syndication seems far removed from the redeemed, now-hip medium that delivers such critical darlings as *The Wire* and *Mad Men*, such popular myths as *Heroes* and *Lost*, and cutting-edge digital experiments like Current TV or ESPN360.

This disregard extends to the media industry itself, where syndication is similarly generally regarded as low-rent and, in the words of at least one market analyst, "unsexy."[3] Such attitudes have long been etched into standard industry practice. Advertising time on syndicated programs is sold last at the annual "upfronts" each spring, after both the broadcast and the cable networks have made their pitches and completed their sales to national advertisers. The programs themselves are relegated to parts of the schedule long described in industry vernacular as "fringe": late weekday mornings and afternoons, late weeknights, and scattered across the weekend—in other words, outside the time that "matters." Moreover, and most tellingly, advertisers, studios, and analysts have historically regarded the presumed viewers of syndicated programs as themselves "fringe": stay-at-home mothers, racial minorities, children, elderly shut-ins, disaffected white men, and low-income households.

However endemic this treatment of syndication, and however much the next-generation spin of new technologies and cultures of media use continually relegates it to the past, the practice persists as a critical component of the US television industry. Indeed, syndication is one of a handful of relatively staid and secure investments in an era of high-stakes, transmedia ventures such as NBC Universal's *Heroes* or potentially lucrative but unruly media sites like Google's YouTube. This chapter explores how syndication has maintained its viability, if not always its dignity, in the face of immense changes in American television. Syndication has remained vital to many corners of the media industry during this period precisely for delivering the one factor lacking virtually everywhere else in the industry: *stability*. Although network and cable prime time has been defined by big-ticket projects, anxious marketing, and the rapid development of multiple distribution platforms, syndication—at its most successful—has delivered consistent, predictable, and diverse audiences daily. The primary reasons for this stability are the programs themselves, which, after a tumultuous wave of bold and often exploitative ventures in the 1990s, have settled into a hegemony of tried-and-true, anti-controversial genres and programs. As media analyst Marc Berman noted in 2004, "syndicators radiate a certain aura of durability in their offerings."[4]

That said, another appropriate word to describe syndicated programs at this time is "banal." There is a comfort, an unremarkable "everydayness" of most syndicated programming, that, even in seemingly outlandish series (e.g., *TMZ*), works to anchor the normative claims of the practice as a fundamental form of television. Banal television is habitual, rather than compelling: "just-see TV" rather than "must-see TV." It is driven not by the staging of dense plots, controversial content, or breathless talent competitions, but by the presentation of a coherent set of low-intensity expectations. During the classic network era such designed banality was television's standard operating procedure; today, syndication is its last major preserve. The fact that the same four daily programs—*Entertainment Tonight*, *Jeopardy!*, *The Oprah Winfrey Show*, and *Wheel of Fortune*—have dominated the syndication ratings for over twenty years is perhaps the most obvious testament to that banality. Moreover, the very form that most syndicated programming takes—live-to-tape, multi-camera studio production—has always been the epitome of "everydayness" on television the world over. While network prime-time and cable networks have regularly squirmed against such normative aesthetic and cultural limits of the medium since at least the early 1980s, syndication embraces the look, feel, and sensibility of "television" in its most conventional forms.

Syndication means "banal television" in another, crucial way: it is watched live, with commercials. The growing shift to delayed viewing in recent years (via DVRs and online sites) has threatened the very rationale of advertiser-supported television. However, the industry trade association SNTA (the Syndicated

Network Television Association), which reorganized in 2003 to bolster the syndication trade, has loudly trumpeted 2007 Nielsen research that claims that 95 percent of 18- to 49-year-old viewers (including 86 percent of DVR users) watch syndicated shows on the day of their broadcast, and 85 percent of DVR viewers watch them live.[5] Moreover, a Magna Global survey from the same period indicated syndicated viewers apparently remember the advertisements they've seen at a higher rate than either network or cable prime-time viewers.[6] If evidence like this is an accurate representation of viewing behavior, then syndication is single-handedly sustaining the traditional model of broadcasting. The SNTA points to plausible reasons for the greater commercial recall, for example shorter commercial pods, but the programming itself must have something to do with these figures. While broadcast and cable networks have increasingly touted "appointment" viewing—in some ways, making the shift to DVR and online viewing a self-fulfilling prophecy—syndication has instead maintained its lower-profile status as "habitual" viewing, i.e., as something audiences just do, rather than plan for. Moreover, as FOX Television Stations CEO Jack Abernathy opined in 2005, "liveness" (or near-liveness, as with day-and-date programs that air on the same day of their production) can be a virtue with particular kinds of programs, "because when you can buy a season of a particular show in DVD and soon that's available online, then the real advantage for television stations is to be live and to be flesh and to be TiVo-proof."[7]

Asked to describe syndication's greatest asset in 2007, SNTA president Mitch Burg modestly claimed, "it's all about trust. Trust and credibility deliver a higher return on investment for advertisers, and no one can provide that better on television than syndicated series."[8] The rapidly changing television environment of the 2000s has certainly challenged that trust, but Burg's claim still resonates. However, the trust that results for advertisers is caused by banality in the programs: the carefully managed consolidation of particular assets, formats, and practices in order to sustain syndicated programming's habitual, steady, everydayness.

Syndication in Transition

The ways in which we make, distribute, and experience media of all forms have changed considerably since the mid-1980s and began changing even more radically in the mid-2000s, i.e., the beginning of the "post-network" era. Syndication, historically poised between networks and local stations, and more reliant on the traditional conception of commercial broadcasting than either networks or cable, has been particularly affected by some of these changes, while remaining surprisingly resilient to some others.

Traditionally, syndication has been the primary mode of non-network distribution of television content from national distributors to local stations. Programming typically offered has included new (i.e., "first-run") and rerun (i.e., "off-network" or, more recently, "off-cable") series, as well as packages of older theatrical films, although this practice has almost entirely ceased due to the overwhelming availability of films on other distribution platforms (e.g., cable, DVD, pay-per-view, and download). Since the late 1980s, syndication transactions have generally been made on a "cash-plus-barter" basis, which means that, in exchange for the right to broadcast particular programs, stations pay some cash but also surrender some advertising time in that program back to the distributor (i.e., "barter" it). The distributor then sells that advertising time to national advertisers. Thus, a system has evolved in which syndication distributors have been able to profit from payments from both stations and advertisers. When a particular series fares well with both of these buyers, it can become a highly valuable broadcast property for years.

Accordingly, the most fundamental change in the organization of television over the past two decades—the consolidation of studios, networks, cable systems, broadcast stations, syndication distributors, and other media entities into a handful of large corporations—has affected the relationship between these sellers and buyers. This market consolidation, fueled by relaxed federal oversight, has affected all media in many ways, as large companies have expanded the range and depth of their holdings, devouring mid-sized and small companies along the way. Although this is a familiar story, found throughout this volume and in most accounts of media industries over the past two decades, the particular effects of this consolidation on the syndication trade have been less understood and more complex.

Since the mid 1980s, four of the largest media conglomerates—CBS, Disney, NBCU, and News Corp.—have acquired not only the production facilities of film studios and the national distribution capacity of national broadcast and cable networks, but also the localized transmission outlets of local television stations in dozens of large and medium-sized markets.[9] Over time, this effectively created a syndication market in which only a handful of corporate buyers—each representing dozens of key stations nationwide—dictate the overall fate of a syndicated program. As recently as the early 1990s, a first-run syndicated program could eke out a marginal existence in a few dozen independently owned stations scattered across the country; today, due to the market control exerted by these gargantuan station groups, such distribution is impossible to achieve, let alone sustain. If the large station groups listed above—collectively controlling 100 stations in most large and mid-sized markets—all decline to purchase a potential series, a syndicator is extremely unlikely to put it into production.

Thus, though the number of on-air stations actually rose during this period, the number of discrete buyers of syndicated programs shrank considerably. "Mid-sized" media corporations—such as Belo, Gannett, Hearst-Argyle, Meredith, Scripps, and Sinclair—which are still quite large by historical standards, each owned a smaller but similarly powerful collection of stations that could contribute the difference in national coverage well before syndicators approached individual station owners. In addition, continued relaxation of federal ownership rules allowed duopolies (the single ownership of two television stations in a market) and local management agreements (the joint management of separately owned stations), each of which further diminished the number of potential buyers of programming. David Garfinkle, the CEO of small syndicator Renegade 83, lamented in 2003 that "four or five years ago there was probably 20 or 25 companies that you could go to and sell to. And now there are the five companies, and you make your five phone calls and you're done."[10]

Matters are just as super-sized on the selling end, as once-independent producers and distributors merged with these same corporations or simply vanished from the market, their existing properties absorbed into their new owners or sold off. For example, the mammoth distributor currently known as CBS Television Distribution (CBSTD) existed as recently as the mid-1990s as several companies: Paramount Television Distribution, Eyemark Entertainment, Worldvision Enterprises, Spelling Entertainment, and King World Productions. As a small number of large station groups controls distributor access to audiences, a small number of large distributors now controls advertiser and station access to programming. In such a market, it is hardly surprising that CBSTD distributed eleven of the top fifteen syndicated series in the 2006–7 season.[11]

This concentration of both ends of the market has benefited corporations such as NBC Universal and CBS Corporation, but not without a cost. As the preferred treatment of network and cable advertising time-selling at the annual upfronts indicates, syndication is given, at best, third priority in media corporations' television business plans. This lack of attention extends to program development at the major distributors that in recent years has generally eschewed experimentation in favor of solidifying established genres (e.g., court shows and celebrity-led talk shows) and zealously supporting long-running hits such as *Jeopardy!* and *The Oprah Winfrey Show*. Although the biggest companies have benefited overall from this strategy (see below), and have indeed grown even bigger in recent years, they have also had to cope with the burden of girth: they may dominate the space of syndication through sheer size, but paradoxically they may not be able to move as effectively as smaller companies might. Many syndication markets, particularly daytime, offer little space for maneuvering. Accordingly, small, "boutique" syndication distributors, such as Debmar–Mercury, Litton, Program Partners, Radar and Trifecta, emerged in the late 2000s to pursue business that "falls through the cracks": too small, too off-beat, too risky, or too Canadian for the major studios to attempt.[12] Hank Cohen, CEO of Trifecta, stated in 2007 that "syndication affords us an opportunity to find our place among distributors as long as we are able to zig when everyone else was zagging."[13]

These small companies have had some success "crack filling," but the overall viable "space" on commercial broadcasting—i.e., the available time slots as well as the likely size of the audience at that time—is still constantly shrinking. If "consolidation" describes the fundamental trajectory of commercial media ownership since the 1980s, "fragmentation" is the usual term associated with media users during this period. Appropriately, this is manifest in many ways today, from the "iPodization" of media consumption via download, to the dispersion of viewers across hundreds of cable, satellite, broadcast, and broadband channels. For syndication, as with network and cable, fragmentation has meant a steady dwindling of audience share. Setting aside the perennially contentious industrial issue of the metrics of ratings systems, the systems themselves—dominated for decades by Nielsen Media Research—still dictate the functioning of the advertising market, which remains the economic engine of commercial television. A rating need only be "high enough" to help sustain a series. The problem is that, while the "floor" for that "high enough" continues to drop, the costs of production continue to rise. Whereas a syndicated show would have been headed for cancellation in the mid-1990s if its rating dropped below a 4 or 3 (meaning 4 percent of homes with television were watching that show), the "high enough" circa 2009 was nearing a rating of 1.[14] This means that CPM rates—the cost to reach 1,000 viewers—continually move upward, giving stations and advertisers pause as they consider whether a particular show is worth their investment. The mixed fates of two key syndicated program forms in recent years—the first-run "action hour" and the rerun off-network series—exemplify how changes in the market structure of the media industries during the transition to a post-network era affected the sustainability of particular kinds of programming.

At the turn of the twenty-first century, which was before the advent of online distribution but well into the era of audience fragmentation, the most expensive type of syndicated programming became the most vulnerable: the weekly first-run "action hours" that anchored most independent stations' weekend schedules throughout the 1990s. These stunt- and effects-heavy ensemble adventure series were designed primarily to appeal to the elusive 18- to 34-year-old male audience and secondarily to anchor entire franchises of ancillary products. The phenomenal success of Paramount's *Star Trek: The Next Generation* in the late 1980s sparked the action-hour trend and inspired most of the major and mid-level distributors to create their own series. The *Star Trek* feature films were performing well overall at this time, and Paramount had been looking for a way to resurrect the franchise as a television series for over a decade. In the mid-1980s, they saw an opening in weekend prime time on independent stations, shrewdly reasoning that the greater control and profit they could gain from this experiment through both license fees charged to stations and the sale of national advertising spots within the program was worth the risk of bypassing the traditional broadcast networks. By the mid-1990s, this "experiment" greatly expanded the *Star Trek* franchise through two additional television series, an extended feature film series, and myriad merchandise. Many other syndicated action hours followed in its wake, including Pearson Television's *Baywatch* (resurrected from a series NBC canceled the year before), Warner Bros.' *Babylon 5*, and Studios USA's *Hercules: The Legendary Journeys* and *Xena:Warrior Princess*, which all met with some industrial, popular, and critical success.

Series such as these were staples of weekend television throughout the 1990s. Mostly formulaic, simply told yarns, they were certainly a significant part of the "banal" television landscape of the time.[15] Their emphases on action and standard "good versus evil" plots also generally sold well overseas, and thus many were produced and distributed by new, ad hoc partnerships of global media capital. As Jay Firestone of the Canadian conglomerate CanWest Entertainment explained, "translating a guy kicking a bad guy out a window is a lot easier than having two cops discuss poverty."[16] However, by 2001, rising production costs, falling ratings, disappearing time slots, a battered economy, and failed launches of increasingly derivative new series effectively squashed further development of this form.[17] Moreover, Saturday night television itself was fast fading, as audience levels dropped to the point that even the Big Four networks essentially abandoned the evening, choosing to fill it with reruns of popular series or feature films. With its biggest successes in the past and its target audiences turning to cable and DVD, the form had seemingly become exhausted. Studios and distributors recognized that the return on investment could no longer warrant the expense of these series and turned their syndication resources toward much cheaper and consequently more lucrative reality shows such as *Blind Date* and *Celebrity Justice*. Several projects that might have succeeded as first-run dramas a few years earlier instead found homes at cable channels such as Sci-Fi (*Farscape*, *Stargate SG-1*) and USA (*The Dead Zone*). These channels evolved and expanded their fare largely through incorporating strategies and genres tested elsewhere, redefining their brands around these series, and amortizing production costs through multiple repeats.[18] Although Tribune Entertainment gamely stuck it out for a few more years with space opera *Gene Roddenberry's Andromeda* and teen superhero drama *Mutant X*, by 2004 the format was essentially dead.[19] The demise of the action hour as a legitimate contender against network fare indicates the practical limits of syndication's boldest and most expensive experiments in a shifting media environment.

Another expensive form of syndicated programming, the off-network rerun, faced similar issues as a result of corporate consolidation and audience fragmentation, but remained an in-demand program form. As I argued in my book *Rerun Nation: How Repeats Invented American Television*, the off-network rerun has long been one of television's most essential program forms. It remains a viable part of syndication, even as the very nature of both "television" and "syndication" becomes less tangible with every new digital distribution venture. Indeed, as

Sharon Sharp notes, even online digital video hubs such as AOL's In2TV and the NBCU-News Corp. venture Hulu are still premised on the comfort and familiarity of particular television shows, exactly the logic of repetition that has driven their valuation in broadcasting, cable, and home video for decades.[20] Thus, while much of local broadcast television is still dependent on the familiarity of the syndicated rerun, rapidly shifting calculations of supply and demand, as well as various layers of "licensed exclusivity," have challenged this relationship.

Off-network situation comedies (sitcoms) are still the favored fringe-time option for local stations. With their half-hour running time and pleasant sensibility, sitcoms have long epitomized "acceptable" television banality—i.e., the kinds of shows that are expected to be habitually viewed over and over again by the same viewers. In the 2000s, *Friends*, *Seinfeld*, and *Everybody Loves Raymond* were far and away the dominant off-network sitcoms, the "crown jewels" locked up in long-term contracts by top station groups.[21] A handful of other 1990s and 2000s sitcoms, such as *Frasier*, *Will & Grace*, *The King of Queens*, and *That 70s Show* also had successful syndication runs, though with lower ratings. After that tier, however, few of the other network sitcoms of the era met with success in syndication, even though station demand for the familiar form remained high. Yet even the "crown jewels" started to lose their luster by 2005 and faced steadily diminishing audiences. Paired with the fact that the creation of new network sitcoms is at an all-time low (in favor of dramas and reality shows), the future of the syndicated off-network sitcom appears more tenuous than it ever has before.[22]

The introduction of Tyler Perry's sitcom *House of Payne* (first syndicated in fall 2008) may, however, auger a new mode of production and distribution for this particular television form. Recognizing the high station demand for popular off-network sitcoms in the mid-2000s, but also aware that the broadcast networks were not a receptive market for his particular style of African-American comedy, Perry self-financed ten test episodes of *House of Payne*. He then partnered with independent syndicator Debmar–Mercury to distribute the series to a handful of large-market independent stations in June 2006. The success of this trial run convinced cable giant TBS to order 90 more, and TBS then scheduled new episodes of *House of Payne* at a torrid biweekly rate, beginning in June 2007. The show scored high ratings and prompted the development of other new sitcoms on the channel. In fall 2008, with over 100 episodes completed in less than two years of production, the series began a syndication run on FOX stations nationwide. While the series is far from stylistically groundbreaking and is cast comfortably in the mold of every post-*Cosby* African-American sitcom, its unusual development—rapidly moving from first run to rerun—points to syndication's continuing ability to distribute material that might not otherwise reach an audience.[23]

Even with its recent difficulties, the off-network half-hour sitcom remains sanctified in syndication; however, the place of off-network drama series—generally referred to without need for explanation as "hours" in industry argot—has never been stable.[24] Syndicated hours have historically been difficult to schedule in daily syndication and have never drawn the size or makeup of audience that sitcoms command. Accordingly, cable networks became the primary home for off-network hours in the 1990s, filling their prime-time schedules with series that would have never run in domestic syndication at that period. As procedural dramas dominated broadcast network prime-time schedules in the 2000s, increasingly aggressive cable networks paid escalating license fees for exclusive rights to reruns of these hits, from the *CSI* and *Law & Order* franchises, as well as their knock-offs (e.g., *Cold Case*, *Medium*, *Without a Trace*, etc.). In most cases, these deals provide them with exclusive rights to these series for a number of years, effectively locking broadcast stations out of contention if they had been interested in the series. However, as the first-run action hours shut down early in the 2000s, the "weekly scripted drama" gap remained for stations to fill on late Saturday and Sunday nights. Thus, weekend runs of popular off-network hours became a standard local programming strategy during the decade, and these shows garnered solid enough ratings to gain renewal and attract local and national advertisers. Moreover, by the end of the 2000s, largely on account of the dearth of viable first-run series and off-network sitcoms, stations were even increasingly willing once more to try reruns of hours as daily strips.[25]

The sharing of popular dramas between cable and broadcast syndication exemplifies a larger trend defining content distribution in the post-network era. Content is no longer exclusively associated with a single delivery platform, but is instead increasingly—and often simultaneously—made available across multiple platforms. Distribution contracts now generally allow stations multiple daily runs of series and even the oxymoronic concept of "shared exclusivity," which allows multiple stations and/or cable networks to run the same series simultaneously. Accordingly, all three of broadcast syndication's top sitcoms of the 2000s—*Friends*, *Seinfeld*, and *Raymond*—continue to run on cable network TBS, have been out on DVD for years, and are beginning to appear online.[26] Thus, as cable, DVD, and online distribution platforms have grown, local stations' role as the exclusive home of off-network reruns has steadily vanished. While they are still generally granted a brief exclusive window for off-network sitcoms, stations now have to wait their turn—or share their distribution windows—with cable networks, DVD box set releases, and various forms of online syndication (i.e., full episodes of the same property available through multiple websites) before gaining access to the more popular dramas.[27] Program distributors have realized the considerable value that can be extracted by licensing multiple windows—with varied terms—to several outlets at once. As John Nagowski, president of CBS Television Distribution, noted in a 2006 interview, such redistribution of existing content is the "low hanging fruit" of syndication's overall strategy in an increasingly online media environment.[28]

Still, despite these diminished expectations, syndication not only weathers on, it prospers. After a dip during the recession of 2001–2, upfront revenue has risen every year since, despite the steadily falling ratings. The difficulties presented above are endemic throughout the media business today, as digital, customizable, on-demand media forms erode the centrality of the traditional broadcast model. So why does syndication, the least sexy of any media form, persist? Syndication's engine has been stability: its ability consistently to deliver particular audiences to advertisers year after year. In addition to the three sitcoms mentioned above, the primary vehicles for this sustained growth have been a handful of established first-run hits, most of which began their runs during the Reagan administration: *Entertainment Tonight*, *The Oprah Winfrey Show*, *Judge Judy*, *The Ellen DeGeneres Show*, *Jeopardy!*, *Wheel of Fortune*, and *Live with Regis and Kelly*. The stalwart dominance of these particular shows is routinely discussed in trade journals, and their stations jealously protect their licenses. As Chris Kager, head of MGM/NBC media sales, expressed it,

> the beauty of syndication is that you know where you're running—you won't be given a new show after six or eight weeks. … [Our] shows are on the air 52 weeks a year and viewers watch them day after day. There's a value to that. In a sea of turmoil, syndication brings you a lot of tranquility.[29]

The Flavors of Banality

Much like the key series of broadcast network history (e.g., *I Love Lucy*, *All in the Family*, *Hill Street Blues*, *Survivor*) the most dominant syndicated series have also popularized and solidified particular forms and genres. In an era of steadily falling ratings, syndication's "tranquility" has been highly valued by stations, advertisers, and distributors alike. Currently, this tranquility is found predominantly in three key first-run genres: the entertainment news magazine; the intimate, lifestyle-focused talk show; and the court show. With the exception of game shows, which remain a consistent facet of syndication, and indeed appear to be growing in popularity at the time of this writing, virtually all of the series in first-run syndication hew to these well-worn formulas. They are the current incarnation of syndicated television banality. Moreover, they have also displayed a great capacity for adaptation to the post-network rules of content distribution and the "diminished expectations" of syndicated success. Aside from the talent costs on star-driven series such as *Ellen DeGeneres*, *Judge Judy*, or *Oprah*, they all have generally low production costs and have taken advantage of alternative forms of program financing (including, most prominently, product integration). In addition, the entertainment and talk shows all have extensive ad-supported websites offering news, blogs, video clips, viewer forums, and additional content that extend what would only be televisual banality into a more interactive form of web banality—i.e., a familiar, comfortable, habitual, but online-appropriate space.[30]

In the remaining pages, I will briefly examine each of these key syndicated genres. Collectively, they illustrate the benefits of banality for syndication at this moment by emphasizing intimacy and the everyday in their content and style. In their rigid adherence to generic norms (save one important exception, *TMZ*), they reinforce the reassuring stability of syndication, the ideological sense that "this is how life is" or should be. As Brad Adgate, a market analyst at Horizon Media, explains, "These are not just individual shows, they're brands."[31] The shows consequently foster a kind of relationship with television rarely found in prime time or on cable. It is a "relationship" more than an "experience" not only because it is everyday and habitual, but also because it solicits direct viewer engagement and increasingly extends to other media: magazines, websites, and mobile media. In the extreme case of *Oprah*, it encompasses a vast range of expressly articulated products, behaviors, and even beliefs with which millions of women identify through host Winfrey. As Mimi White argues in her analysis of television's therapeutic discourses, this relationship is expressed and understood as "given" by these programs and their viewers: "[The] idea that television functions as a therapeutic apparatus, and should be explored in these terms, is an integral part of the everyday discourses and practices of regular television viewers."[32]

Entertainment News Magazines

At first glance, entertainment news magazines may seem unlikely to forge the sorts of relationships with viewers that lifestyle and even court shows proffer. After all, unlike the other genres examined here, entertainment news shows focus not on the everyday lives of "real people" (or at least their aspirational ciphers), but on the decidedly "unreal" lives of celebrities. However, *pace* White, the therapeutic function of entertainment news as a daily fixture is to merge the "reality" of displayed celebrity lives into the everyday cultural texture of "regular" viewers. Celebrity media coverage has long been a source of daily conversation and speculation (i.e., as "water-cooler" talk), with celebrity lives offered as a seemingly personal connection to media culture. Celebrity news and gossip is particularly pitched at young adults and has become a prominent feature across all media on account of the longstanding belief that young people are more reliable consumers and thus a more attractive proposition for advertisers, who still traditionally favor marketing to young adults. Accordingly, advertisers and media firms have capitalized on this deepening relationship between young adults and celebrity news. As Jane Buckingham of marketing firm Youth Intelligence said of the syndicated entertainment news magazine format: "You can get a pop culture roundup with little commitment of brain power or time. It's escapist, yet it gives you a knowledge currency."[33] Thus, as with most areas of news coverage by the 2000s, this programming places a premium on delivering current cultural capital in as small a package as possible.

Celebrity gossip had been a fixture of Hollywood news coverage for decades before *Entertainment Tonight* premiered in 1981. That said, the program brought a quasi-journalistic and highly televisual focus to Hollywood publicity by adapting the tactics of 1970s ENG (Electronic News Gathering) with on-location coverage, fast-paced video editing, and a "day-and-date" production schedule that enabled episodes to be recorded in the morning and broadcast in the early evening of the same day. As with the "action news" model, this insured the timeliness of coverage and was designed to propel habitual viewing—i.e., the series' constant teases of what was "coming up" in future segments compelled future viewing. Viewers *had to know* what was happening, even it was only a film premiere or a behind-the-scenes look at a pop star's latest video.

Since that time, other entertainment news magazines have been attempted and some have become as habitual as *Entertainment Tonight*. In 2008, *ET* competed with three daily syndicated entertainment strips: *Access Hollywood* and *Extra*, which both started in the mid-1990s, and Gen-Y upstart *TMZ*, which is based on the notorious and highly popular gossip website and premiered on television in 2007.[34] The burgeoning appetite for celebrity gossip in recent years across all media derives largely from these series, most of which operate in similar fashion, quickly covering publicity events, premieres, reactions to scandals, and the like with a combination of glossy in-studio anchors and field reporters. Most of these programs present the celebrity world as a melodramatic spectrum of emotions: breathlessly exciting and glamorous in the "good times" and cloyingly sentimental or vicariously scandalized in the bad.

On *TMZ*, by contrast, celebrities are covered with a kind of cynical, irreverent, and insatiable curiosity. Based on the extremely popular gossip website of the same name, *TMZ* offers celebrity news stripped down to mere celebrity presence: the only "stories" it covers are random celebrity sightings captured on video. Hounded by video cameras, and ever unkempt, the celebrities of *TMZ* are thus seemingly cut down to a "real" scale. In the glare of candid coverage, grime replaces glitz. Instead of walking on the red carpet, they are seen hustling into their cars, walking their dogs at dawn, or hiding behind handbags as they walk through airports. On- and off-screen commentators assess celebrities' appearance and demeanor in these awkward situations, setting up the terms for viewer engagement. For example, an August 2008 segment on country singer Billy Ray Cyrus pointed out how he consistently tried to strike nonchalant poses as soon as he realized cameras were present. Accordingly, unlike the other entertainment series, *TMZ* focuses on leering pictures and video rather than on the hosts and reporters. Instead of a sleek broadcast studio and model-perfect interviewers, *TMZ* displays shaky hand-held shots of a large, bland office full of white cubicles and cute (though far from glossy) young data miners (the staff of the *TMZ* website), all presided over by their decidedly unglossy and unsentimental managing editor, Harvey Levin.

Virtually identical to its website and those of its competitors (primarily PerezHilton.com), the televised *TMZ* has succeeded as syndicated fare by taking the idea of habitual celebrity news down to its core. Produced on a shoestring budget and thus highly profitable, it proves that it's not the public events and glamour of celebrity that matter most to its viewers, but the sheer fact of mediated celebrity itself: i.e., the camera and/or microphone bearing witness to the unscripted presence of the famous. Its success also indicates the changing metrics of syndication; while it consistently draws a smaller audience than the other entertainment magazines, it fares much better with young adults (and young men in particular). Recognizing this loss of younger viewers, the older news magazines have begun to incorporate some of the tactics of *TMZ* and the more web-centric celebrity sites, such as shortening segments and focusing more on gossip than standard publicity, as well as by extending their own websites and services.[35] Since entertainment news magazines depend on a young audience, such changes are necessarily for survival, though ultimately only update the series' comfortable banality for the Web 2.0 generation.

Lifestyle Talk Shows

Since the explosion of so-called tabloid television in the 1990s, daytime talk shows have been one of the most visible and studied genres on television. They function as everyday, accessible arenas for emotional debate about issues that strike close to the supposed core of American middle-class identity: sexuality, religion, race, gender, and class. Despite their high audience levels, however, their notoriety among cultural and religious tastemakers in the early 1990s probably contributed more than anything else to syndication's down-market reputation in the media industry and popular culture. At that time, their highly public spectacles of personal and social conflict, as well as their unprecedented—and often contentious—use of subjects and studio audiences from outside the hegemonic white heterosexual middle class, drew the attention of many media scholars, who, through textual, industrial, and audience analyses, cogently located the genre in the context of key late twentieth-century cultural politics.[36] However, after a sustained string of public and industrial criticism and several violent and even fatal incidents involving talk show guests, the genre as a whole attempted to upgrade its public image beginning in the late 1990s.[37] This shift, which largely dispensed with the most controversial discussion topics and individual subjects (particularly on the top-rated shows), has made the talk show more palatable to advertisers and has reinforced the association of the genre with contemporary American middle-class femininity. In other words, its particular brand of banality has quietly moved upscale.

In making programs more "palatable," talk show producers have redefined the genre's function away from the controversy of socially marginal figures and issues and towards the definition of the ideal self—i.e., looking "inward" rather than looking "outward." Thus, their primary function has become lifestyle—presented in talk shows as rational, self-generated choices of behavior and consumption—rather than social engagement. No program better exemplifies this philosophy than *The Oprah Winfrey Show*—long the top-rated and most lucrative talk show on television (syndicated or otherwise). The motto of the series and its many ancillary products is "Live your Best Life," a commandment to embrace the idea of lifestyle as *the* means of obtaining success and happiness. In recent years, this mission has prioritized such markers of upper middle-class success as literacy, financial management, and mental health, as well as more consumer-oriented topics such as fashion and technology. In addition, the program has always taken on particular social causes, including racism, the healthcare crisis, and endemic poverty. Significantly, Winfrey herself is at the core of the series' mission; the "best life" the show advocates is personified in her choices, beliefs, behaviors, and (not least) body. While there is certainly nothing banal about the social goals the program endorses or the benevolent effect Winfrey has no doubt had upon them, their framing and treatment as daily talk show fodder generally limits the scope of action to individual choice. Moreover, in the occasional presentation of more socially complex issues—such as a 2008 segment in which panelists including documentarian Michael Moore debated the healthcare crisis—the show's sensibilities rarely extend beyond revealing great concern about the topic, thus limiting potential viewer action beyond experiencing emotions such as "outrage" or "hope."

Other successful series cover similar issues from a similar individualistic perspective, including most prominently *Oprah* spinoff *Dr. Phil*, which features the psychological advice of family counselor Phil McGraw, a former weekly guest on *Oprah*.[38] Industrially, these shows have almost entirely supplanted *Springer*-styled programs as a result of the continued success of *Oprah* as an entire lifestyle orientation built around a multi-media brand that includes a magazine, book club, cable channel, and television production company. The lifestyle shows have been much more attractive to advertisers than the series of the 1990s, as they have successfully nudged the talk show audience into more affluent demographics, created opportunities for similarly positioned daytime lifestyle shows (e.g., the home-focused Rachael Ray and Martha Stewart shows), and reinforced the genre's industrial rhetoric of "stability."[39] As with the earlier talk shows, audience engagement (both in the studio and at home) has been a primary factor in a series' industrial, popular, and critical success or failure. The key difference now is that that audience is—for the most part—anchored more firmly in hegemonic middle-class goals and oriented toward individual actions and acquisitions rather than social engagement, as in the case of a show such as *Donahue* or even earlier versions of *Oprah*.

At the same time the issue-based talk show has become lifestyle-focused, a second subgenre of talk show that focuses more on entertainment than on encouragement has become a staple form of syndicated banality. Instead of lifestyle being made a project for viewers, here it is implied by the comfortable setting, witty, always entertaining host, and constant stream of equally pleasant celebrity guests. Fueled first by the success of *The Rosie O'Donnell Show* in the late 1990s, and most recently by *The Ellen DeGeneres Show*, these series function as more feminine takes on the David Letterman model of late-night talk shows: mildly irreverent and never controversial. As Elizabeth Herbst-Brady, a media buyer at Starcom USA, describes: "A show like *Ellen* is what can be called a 'happy show.' That's a show where people are talking and people are laughing and someone is singing a song, and the home audience is considered more willing to absorb the advertiser's message."[40] More to the point, Michael Teicher, an executive at Warner Bros. Domestic Television Distribution, the syndicator of *The Ellen DeGeneres Show*, cited the series' attraction to "a female audience with lots of disposable income."[41] Additionally, like the late-night talk shows and entertainment news magazines, these series are generally recorded day-and-date, allowing them to comment on current events, and, given their "light and happy" perspective, further reinforce the cultural capital of celebrity gossip.

While these formats have successfully claimed the upper end of the daytime demographic, they are not invulnerable. Many series died attempting to parrot these formats, including a string of conspicuously expensive failures hosted by celebrities such as Tony Danza, Megan Mullally, and Jane Pauley. Moreover, with the most successful of these series booked on stations for the foreseeable future, and with the available attention of the desired demographic arguably at its limits, there seems to be little space in the schedule left for new entries in this subgenre.

Court Shows

Both the entertainment news magazine and the lifestyle talk show produce discrete versions of aspirational everydayness, but the court show genre—more than any other currently in syndication—epitomizes the banality of syndicated programming itself. Court shows claim more daytime space on television schedules across a broader swath of stations than either of the other formats and are increasingly regarded as the safest bet: i.e., the format most conducive to the current market conditions. Production costs account for much of this calculation. "Judge Judy" Judith Sheindlin's massive salary notwithstanding, court shows are arguably the cheapest format on contemporary American television. The shows require a minimal set, minimal paid on-screen talent, and minimal pre- and post-production; an entire 39-week season is typically shot in less than two months production time. More than cost considerations, however, court shows are also increasingly valued for their sheer efficiency as compelling television that combines many of the elements that propel other syndicated forms such as the lifestyle talk show and the off-network rerun. As Jim Paratore of Warner Bros. Domestic Television Distribution described in 2005, "[the court show is] a simple, repeatable format. It has conflict and resolution in a tight package, and if you have a central host that's compelling and authentic, it all comes together into something that is pretty formulaic that works."[42] Moreover, while all syndicated formats are steadily losing audiences, court shows have the slowest rate of viewer erosion.[43] Accordingly, by the end of the 2000s, the number of court shows in syndication had, for the first time, equaled the number of talk shows.

Underneath that industrial logic, however, they are also, in many ways, the most socially significant genre in syndication—if not on all television. The particular banality they offer displays a model of American society wherein authoritative women or men dispense rapid, direct judgment upon the relatively low-stakes interpersonal disputes of people from the margins of normative (i.e., hegemonic middle-class) society. Each series makes a clear claim for authenticity, as text and voiceovers remind viewers that the cases, litigants, and outcomes "are real." In exchange for streamlining the process (and likely sacrificing some legal rights), litigants surrender their fates to the media apparatus and experience a justice system ruled by the conventions of television drama and the personality of the presiding television judge. As Laurie Ouellette argues in her analysis of *Judge Judy*, the dominant template upon which the format has evolved, court shows presume the hegemony of a neoliberal social order in which individuals, and not society, are solely responsible for their lot in life.[44] Accordingly, given that orientation and the extreme diversity of court shows' litigants and audiences (on screen and at home), it is socially appropriate that many of the television judges are women of color. Their very presence, as powerful figures dispensing official justice, enhances their interaction with litigants, and, more significantly, underscores the neoliberal message of self-responsibility. Program developers, acutely aware of such associations, highlight them in their pitches to stations and advertisers. Touting the authenticity of *Family Court with Judge Penny*, Josh Rafaelson of Program Partners claimed Judge Penny Brown Reynolds's "upbringing in an impoverished single-parent home has uniquely prepared her for this series."[45] Similarly, Judge Cristina Perez's five-year run on a Spanish-language court show on Telemundo was regarded by her distributor as an important factor in consistently attracting Hispanic viewers to her English-language syndicated show *Cristina's Court*.[46]

As with all the other genres, however, there is only so much room in the marketplace for virtually identical series, and the shows place a high premium on the judge's distinct personality. Given Sheindlin's dominance, many stations have opted to double run (or more) her series instead of purchasing an additional court series.[47] Accordingly, some court shows are experimenting with techniques that extend the visual range of the format and further capitalize on their personalities' interpersonal skills to guard against format exhaustion. For example, *Street Court*, a series in development as of this writing, will abandon the courtroom entirely and instead dispense its justice *in situ*, at the locations of the disputes.

Conclusion: Syndication Survives

Despite ever-attenuating ratings, television stations' appetite for programming continues. They still must sell advertising to survive, and syndication remains the best available option for many hours of the day. Its key genres are steady and familiar, and distributors are finding ways to expand their presence in new media environments, such as through simultaneous windows and web communities. Key programs *Entertainment Tonight*, *The Oprah Winfrey Show*, and *Judge Judy*, as well as game shows *Jeopardy!* and *Wheel of Fortune*, continue to dominate ratings and have been renewed well into the 2010s. However, given the age of these shows and their personalities, speculation continues to emerge about how long their dominance can be maintained and whether—and to whom—their respective "torches" can be "passed." While discourses of stability, particularly from its main trade organization, SNTA, still dominate perceptions of syndication, independent developers and distributors have successfully, and increasingly, found corners of television the market has heretofore ignored. Firms such as Debmar–Mercury have recognized that syndication can still be more than talk and court shows and that alternative forms of development can yield successes, as in the case of *House of Payne*. In addition, "off-cable" runs of series once thought incompatible with broadcast syndication on account of their risqué content and more "narrowcast" appeal, shows such as *Sex and the City*, *South Park*, and *Punk'd*, have been successfully introduced (in "cleaner" form) into local broadcast schedules by smaller distributors. In recent years, stations have even brought back ancient off-network series from as far back as the 1960s, helping them fill their schedules economically.[48]

As in every corner of media culture today, however, the Internet still represents the primary incarnation of "the future" as a rapidly evolving space of content distribution, interactivity, and advertising. Syndicated programs have expanded their brand through extensive sites and online content syndication, and websites such as TMZ.com are increasingly being used as "incubators" for new properties.[49] While syndicators and advertisers envision these new opportunities in online syndication as the next logical step in distribution, broadcast stations wonder what role they'll be able to play in this developing market. Some mid-sized station groups are putting more resources into local programming and are hoping to develop projects that can even be syndicated nationally, independent of the major distributors, to offset fears of coming irrelevance and keep their local stations viable.[50] In addition, as broadcast ratings run the gamut from "flat" to "down," stations are increasingly expanding and marketing their own websites, as online advertising sales account for a growing proportion of their overall revenue.

Despite dwindling schedule "real estate" and those "diminished expectations," major syndicators keep offering new first-run series, building upon and minimally tinkering current formulas of banality and hoping for the next Judith Sheindlin or Oprah Winfrey to provide long-term stability. Accordingly, while syndication will likely remain nobody's favorite form, it will also likely remain a critical component of the business of television.

Notes

1 SNTA president Gene DeWitt, Quoted in Steve McClellan, "DeWitt on Selling Syndication," *Broadcasting & Cable*, February 18, 2002, 10.
2 Daisy Whitney, "Rachael Ray Sauces up Syndication," *Advertising Age*, May 8, 2006, S16–S24.
3 Elizabeth Herbst-Brady of Starcom USA, Quoted ibid.
4 Marc Berman, "Familiar Faces," *Brandweek*, September 27, 2004, S10.
5 Chris Pursell, "Delivering Trust in a Syndie Market," *Television Week,* February 26, 2007, S7; Paige Albiniak, "Syndicators Upbeat About Upfront," *Variety*, May 5, 2008, 16.
6 Chris Pursell, "Studies Prove Syndie Strength," *Television Week*, April 16, 2007, 38.
7 "Daytime Dogfight," *Television Week*, December 12, 2005, 12–16.
8 Pursell, "Delivering Trust in a Syndie Market," S7.
9 The largest media conglomerate, Time Warner, owns a considerable swath of motion-picture and cable television production, distribution, and transmission entities, but does not currently own any broadcast television stations.
10 "Changing Definitions of Syndie Success," *Television Week*, December 15, 2003, 25–30.
11 The remaining four were distributed by Sony Pictures Television, Warner Bros. Television Distribution, and Disney–ABC Television Distribution.
12 James Hibberd, "Betting on a Syndie Trifecta," *Television Week*, October 30, 2006, 1–41. Program Partners syndicates weekly reruns of Canadian genre series such as *Da Vinci's Inquest* and *ReGenesis*.
13 Chris Pursell, "Start-Up Saw Void in Syndie Market," *Television Week*, January 2007, 13–15.
14 Some syndicators recognized this trend early on and adjusted accordingly. As long-time comedy and talk show syndicator Byron Allen stated in 2001, "we're no longer in a 2.0 rating business; we're in a 0.7 business." Chris Pursell, "Picture gets smaller for syndication," *Electronic Media*, 20, February 12, 2001, 1.
15 That said, I don't mean to suggest that all of these series limited their dramatic range to habitual viewing. Each of the series mentioned, while certainly heavy on action and effects, experimented with tone and style and pioneered the sorts of complex characterizations and narrative arcs that have since become standard on "quality" television drama. *Xena*, in particular, while leading the action-hour ratings throughout the late 1990s, became a global television hit in large part due to its unique mixture of myth, action, comedy, and pathos, and was also hailed by queer and feminist scholars as depicting one of the richest lesbian relationships in television history. See Sara Gwenllian-Jones, "Histories, Fictions and *Xena:Warrior Princess*," *Journal of Television and New Media*, 1 (December 2000), 403–18.
16 Daniel Frankel, "Tribune-Fireworks Leads Action Hours," *Variety*, January 21, 2002, A3.
17 Daniel Frankel, "Action Hours Inactive," *MediaWeek*, March 5, 2001, 7.
18 For more on cable network strategy during the late 1990s and early 2000s, see Derek Kompare, *Rerun Nation: How Repeats Invented American Television* (New York: Routledge, 2005), 184–90.

19 John Dempsey, "Syndication Doesn't Get Traction on Action," *Variety*, August 2, 2004, 17. In November 2008, Disney distributed a syndicated fantasy action hour, *Legend of the Seeker*, from veteran genre producer Sam Raimi (*Hercules*, *Xena*). Although the ratings in its first season were respectable, pulling in between 2 and 3 million viewers each episode, it remains to be seen whether the series can inspire the revival of the syndicated action hour. Tribune Entertainment itself gradually folded its tent throughout the remainder of the 1990s and is currently dormant as new Tribune owner Sam Zell determines what do with its assets.

20 Sharon Sharp, "Televisual Time Travels: Industrial Nostalgia and the Migration of Television," paper presented at Console-ing Passions: The International Conference of Feminism, Television, New Media, and Audio, Santa Barbara, California, April 24–6, 2008.

21 *Friends* has run on Tribune stations, while both *Seinfeld* and *Raymond* have run on FOX stations.

22 In 2003, Sony Pictures Television executive Steve Mosko noted a trend toward more "front end" payoffs in network planning, i.e., programming series that can generate large ratings and revenue on their initial network runs, such as *American Idol*. Calling this development "realistic," Mosko argued that networks and studios "know the backend may not be as big, so they get it where they can—and they get it now." Still, the successful addition of *Two and a Half Men*, *Family Guy*, and *George Lopez* to syndication in fall 2007 seems to have temporarily calmed fears about a shortage of viable sitcom properties. Michael Schneider, "Off-Net Riches No Longer Guaranteed," *Variety*, September 15, 2003, 15.

23 See A. J. Frutkin, "Payne Breaks the Mold," *MediaWeek*, May 28, 2007, 10.

24 See Kompare, *Rerun Nation*, 184–90, for more on off-network dramas at this time.

25 One independent distributor, Litton, even marketed reruns of weekly 1990s staple *Baywatch* as a retro nightly strip.

26 As of this writing, clips of each series are available online through various sites, including Hulu, TBS, and The WB.com. Full seasons of *Friends* are also sold on the iTunes store. Other venues and formats are sure to follow soon, for alll three series.

27 For an early look at the erosion of exclusivity, see Jim McConville, "Peeking at the Future Through a Shared Window," *Electronic Media*, January 8, 2001, 31. Cable networks, in particular, have successfully wooed advertisers and distributors to their platforms with more flexibility in advertising placement, number of runs, on-screen promotions, and bundling with online platforms than local stations have typically been able to offer. See John Consoli, "Cable Will Target Off-Net Syndication," *MediaWeek*, April 11, 2005, 8–9.

28 "Syndie Leaders Mining New Methods Together," *Television Week*, December 11, 2006, 36–42.

29 Eric Schmuckler, "Pitching the Produce," *MediaWeek*, March 8, 2004, 36–42.

30 Paige Albiniak, "Building the Next Syndication Hit," *Broadcasting & Cable*, August 18, 2008.

31 Berman, "Familiar Faces."

32 Mimi White, *Tele-Advising: Therapeutic Discourse in American Television* (Chapel Hill: University of North Carolina Press, 1992), 27.

33 T. L. Stanley, "Entertainment Programs Have a Heady Following," *Advertising Age*, March 7, 2005, S16-S17.

34 In addition, daily *ET* spinoff *The Insider* also stakes similar ground of its parent, though with an ostensibly tighter focus on "the dirt."

35 In 2008, *Extra* started a cross-promotion partnership with the gossip site NGTV.com to feature its irreverent interviewer Carrie Keagan ("Barbara Walters on Red Bull and acid," according to *Extra* producer Lisa Gregorisch-Dempsey) on the television series. Ironically, the website's acronym stands for "No Good TV." Daisy Whitney, "'Extra' Turns to Web," *Television Week*, December 10, 2007, 1–23.

36 See, in particular, Kevin Glynn, *Tabloid Culture: Trash Taste, Popular Power and the Transformation of American Television* (Durham, NC: Duke University Press, 2000); Jane Shattuc, *The Talking Cure: TV Talk Shows and Women* (New York: Routledge, 1997); Julie Engel Manga, *Talking Trash: The Cultural Politics of Daytime TV Talk Shows* (New York: New York University Press, 2003); and Laura Grindstaff, *The Money Shot: Trash, Class and the Making of TV Talk Shows* (Chicago: University of Chicago Press, 2002).

37 Only one continuing series, *The Jerry Springer Show*, still maintains its trademark 1990s style.

38 Personalities, books, and concepts featured on *The Oprah Winfrey Show* have emerged as a burgeoning subgenre in itself. In addition to *Dr. Phil*, the health-related series *Dr. Oz* and *The Doctors* were both derived from appearances on *Oprah*. The national advice best-seller *The Secret* was propelled to series development from multiple syndicators based largely on positive and extensive coverage on *Oprah*.
39 Kate Fitzgerald, "Crowded Talk Circuit Heads Upscale," *Advertising Age*, May 12, 2003, S12.
40 Lee Alan Hill, "Content a Factor in Syndie Choices," *Television Week*, March 21, 2005, 13–16.
41 Ibid.
42 "Daytime Dogfight."
43 Paige Albiniak, "Superior Court Shows," *Variety*, March 31, 2008, 26.
44 Laurie Ouellette, "'Take Responsibility for Yourself': Judge Judy and the Neoliberal Citizen," in *Reality TV: Remaking Television Culture*, ed. Susan Murray and Laurie Ouellette (New York: New York University Press, 2004), 231–50.
45 Marc Berman, "The Show Will Go On," *MediaWeek*, January 28, 2008, 14–20.
46 Jeff Zbar, "Law and Disorder Are Hot," *Advertising Age*, March 27, 2006, S8.
47 The news that *Judge Judy* was the sole first-run syndicated series, of any genre, actually to gain viewers in the 2007–8 season prompted CBS Television Distribution immediately to extend Sheindlin's contract to 2013 and sell contract extensions to stations.
48 For example, since 2005, the Little Rock-based Retro Television Network has syndicated packages of over 50 different off-network series—most of which are available otherwise online, on cable, and on DVD—from the 1960s through the 1980s to dozens of medium- and small-market stations across the country.
49 Chris Pursell, "Syndicators Take Stock: When a 1.0 Is a Winner," *Television Week*, September 17, 2007, 30–31.
50 Katy Bachman, "Independent Producers," *Media Week*, May 21, 2007, 18–20.

Chapter 4

Home is Where the Brand Is

Children's Television in a Post-Network Era

Sarah Banet-Weiser

My seven-year-old daughter considers herself a fan of the Nickelodeon television programs *Drake and Josh*, *Zoey 101*, and, especially, *iCarly*. While she occasionally watches these programs on an actual television set, much more often she watches them as streaming videos on the Nickelodeon website. When I asked her why she preferred the computer to the television, she replied that she liked to have choices as to what episodes she watched; sometimes she watches a single episode repeatedly. She also replied that she likes to switch back and forth to another website she frequently visits, Webkinz, an online world where kids own both a real (a stuffed toy) and a digital version of a pet for virtual interaction. The Nickelodeon website she visits has links to television programs, but these are not the most prominently displayed images on the site: it also has links to video games, shopping, movie sites, "buzz" (gossip about Nickelodeon stars and programs), the NickCruise (a themed cruise line targeted toward families), Nickelodeon programs, and "Nicktropolis," which is a DIY-themed link, with message boards and opportunities to upload one's videos, create mash-ups, and build one's own website. Additionally, the site is covered with advertising for both Nickelodeon merchandise as well as other toys and games. Evidently, Nickelodeon is not simply a television channel.

I begin this chapter with this anecdotal foray into the contemporary world of kids' television because it vividly illustrates some of the dynamics of the contemporary children's post-network television environment that I subsequently explore in theoretical and industrial terms. Clearly, what it means to be a "fan" in the current environment has shifted from a simple loyalty to a particular program. Rather, the norm of media consumption for youth indicates an investment in a multi-media atmosphere, a particular level of technological acuity, and an affinity with brand culture. When the field of media studies was emerging in the last decades of the twentieth century, scholars such as Stuart Hall and David Morley reminded us that television watching is a "social practice."[1] The insistence that television watching was a *practice*, a kind of social activity, was meant at the time as a challenge to previous approaches to television watching, primarily those methods that measured the "effects" of television as well as critical theoretical work that assumed that audiences were "cultural dupes," passively subject to the ideologies and dominant messages of media corporations and culture industries. However, in the contemporary era of post-network television, "new" technologies, increased access to online activity in the form of user-generated content, and the routinization of multi-media formats have made the "social practice" of watching television even more complex. In particular, the blurring of the distinction between consumers and producers and the ballooning activity of users on sites such as YouTube (where ostensibly anyone with video and uploading capabilities can have their "own" television channel) have forced us to rethink the subject position of the audience. This is as true for children as for adults, and perhaps even more so, given that recent generations have more technological savvy and skill than previous ones.

In this chapter, I explore this historical dynamic as a way to offer some thoughts about the contemporary context of children's television. This history is understood through an economic lens, because the relation between children and media (as children's television extends beyond the television in its post-network era) is always situated within the commercial context that both creates and sustains contemporary media. This history also provides us with an important backdrop in which to examine contemporary media practices, where we can no longer simply regard only television when examining "kids' television." That is, the space of children's television has broadened in the contemporary era to include multi-media platforms that engage kids in websites and ancillary products, but also in practices of production of media by way of interactive media and user-generated content. In this regard, I first offer a brief history of commercial broadcast television for children in the United States. Children's programming in the early days of television reflected ambivalence on the part of the networks, as they were required by the Federal Communications Commission (FCC) to air some programming for the child audience; however, they did not recognize children as an ideal commercial market until the widespread deregulation of the communications industries in the 1980s. The next section presents some of the changes in kids' television and viewing practices resulting from the advent of cable and the normalization of niche markets that led to the cable industry's experimentation with dedicating entire channels to children's programming. This risk, at least for corporate powers such as Nickelodeon and Disney, paid off, and kids began to be recognized as a powerful, multi-dimensional market.

Finally, in the third section, I offer some thoughts about the recent mediascape for children in which the "social practice" of watching television is but one factor, and not necessarily the central element, in a media- and technology-rich milieu. What it means to be a child "fan" of television began shifting in the cable era and continues to adjust into the contemporary new media context. In some ways, the shift has been one of intensity, where children are marketed to in an unprecedented way, on an ever-increasing number of media platforms. In other ways, the shift in the child audience has been a more substantive one, where youth subjectivity—as both a product and a reflection of media—has changed so that older paradigms of media analysis no longer make the same sort of sense as a way to understand the complex relationship between youth culture and television. Because the transformations in the contemporary youth media landscape are multi-dimensional (and thus not simply economic strategies), examining what kids' television "means"—economically, politically, culturally—requires a revised interpretive lens. Situated within brand culture, with opportunities to interact with new media and to become particular kinds of producers of media, kids can no longer be thought of as mere consumers when it comes to television and other media.

Kids' TV: A Brief History of the Network Age

Children's television has been a central element in broadcasting since the emergence of television in the United States in the late 1940s. As Lynn Spigel and others have documented, the appearance of television in the late 1940s and early 1950s as a staple of the white, middle-class American home required a marketing apparatus that would convince consumers of the necessity of this new appliance.[2] Part of this marketing apparatus was then, as it is today, an appeal to the synergy between the television and family togetherness—the television was marketed as a replacement for the family hearth, as Spigel argues, as a way to bring the family closer together. Of course, a significant element in this family togetherness was the child, so television from its inception in the United States was a technology and social practice invested in providing programming appropriate for the child viewer. This investment, however, was somewhat at odds with the economy of the television industry: as for-profit, privately owned institutions, television networks could not rely on the child viewer directly to provide revenue.

Thus, in the early days of television, children were initially recognized as a potentially important market, but mostly in terms of the degree to which they could drive parents to purchase a television set rather than as a group targeted by sponsors or advertisers through programming in the manner familiar today. By the 1950s, as Donna Mitroff and Rebecca Herr Stephenson point out, "Programmers, advertisers, and networks quickly learned that children often influenced parents' decision to purchase a television; therefore, providing programming for children became a priority in the early years of television."[3] And, even in these early days of television, the power of the child audience wasn't limited entirely to influencing parents to buy a television set—the TV, according to Spigel, was situated almost as a kind of parenting tool, something that could help bring unity and cohesion to the family circle. Early broadcast television carried a promise of smoothing social tensions through a commercial venue; as Spigel writes about television in post-war US culture, "television was depicted as a panacea for the broken homes and hearts of wartime life; not only was it shown to restore faith in family togetherness, but as the most sought-after appliance for sale in postwar America, it also renewed faith in the splendors of consumer capitalism."[4] However, while children were certainly an influence market with their parents (meaning that children were now perceived to have a powerful influence on the purchases their parents made), programming tended to be directed more toward "family" viewing than toward a discrete child audience.

Much of early children's programming was thus not actually limited to particular blocks of time that were considered most attractive to kids; rather, many programs were aired in the evenings and were seen more as appealing to a broader audience.[5] This practice, however, was tempered by some negative attitudes toward television. As with all communication technologies, there was a general ambivalence on the part of viewers, legislators, and educators as to what television had to offer the American public. The content of television was seen, as it would be for years to come, as potentially damaging to children's well-being as well as a threat to parental authority. Television networks quickly capitalized on this anxiety and worked efficiently to shape television programming as a cultural product that could be a helpful accompaniment to parenting rather than a detriment.

Indeed, networks had to contend with heightened issues of regulation during the first decades of television in the United States, especially regarding images of violence and other "offensive" themes within programming. Not surprisingly, regulation of such images often centered on children and what was considered appropriate fare for this audience. As Spigel, Heather Hendershot, and others have documented, children's television has always been subject to regulation, but frequently in contradictory ways that depended on both the economic and political climate at the time.[6] While in the 1960s and 1970s the FCC lightly regulated children's television (spurred on in large part by then FCC chairman Newton Minnow's famous "vast wasteland" speech), it was also becoming clearer that advertisers were the dominant voice in determining television content.[7]

By the 1970s and 1980s, then, with the rapid growth of the entertainment industry, audiences began to be more and more parceled out as particular demographics to which advertisers targeted their wares. Despite the increasing market segmentation, children, though seen as a discrete audience, were still not recognized as an entirely profitable one, which resulted in what Jason Mittell calls the "Saturday morning exile": he notes,

> prime time air increased in value and children's shows, primarily cartoons, were relegated to Saturday morning because of their inability to attract advertisers willing to pay increased rates. Within the confines of Saturday morning, networks aired cartoons that had been produced for theatrical release and recycled for television broadcast—a great departure from the prime time programming of earlier years.[8]

The move to Saturday morning, then, effectively designated cartoons as primarily kids' fare, thus simultaneously creating and supporting the belief of the child audience as in need of protection from adult material. Cartoons were regarded as "harmless entertainment" and the narratives and creative style of cartoons were shaped to fit the (perceived) unsophisticated child audience.[9]

At the same time, however, the merchandise tie-ins with children's television increased, and more and more advertisers began to tap into the market potential of the child audience over the course of the decades between the 1960s and the 1980s. With the widespread deregulation of the communications industries during the 1980s under the Reagan administration, an FCC restriction about what it considered "program-length commercials" was lifted to allow for a different kind of "public interest" broadcasting, where the "free market" ethos of the communications industry considered all members of the audience, including both young children and advertisers, as equal kinds of consumers.[10] This led to a boom in the merchandising tie-in business, one that continues in the ever-far-reaching scope and breadth in the contemporary environment. As Tashjian and Naidoo state:

> From toys, clothing, books, and video games to toothbrushes, bubble bath, cereal, and snack food, store shelves are overflowing with products prominently featuring children's television and cartoon characters ... This character invasion is not a coincidence. By June 2004, just 4 years after the character's product launch, SpongeBob SquarePants products had generated more than $3 billion in retail sales of merchandise ranging from bandages, blankets, and bottles to toys, toilet tissue, and t-shirts. As retailers had strategically planned, SpongeBob could be found anywhere and on most any consumer product imaginable.[11]

The increasing scope of kids' television—moving from programming to a complex relationship between television, film, toys, and other merchandise—was an important moment of what now is an even broader and more ubiquitous kids' media culture. The commercial tie-ins prompted yet another FCC intervention several years later with the 1990 Children's Television Act (CTA). Enacted by Congress on October 18, 1990, the CTA was intended to increase the quantity of educational and informational broadcast television programming for children and to force broadcasters to serve the child audience as part of the obligation to the public interest. In 1997, the FCC modified the CTA to include the Three-Hour Rule, mandating that commercial broadcast stations air a minimum of three hours a week of programming that fulfills the educational and informational programming directive.[12]

However, these regulatory measures did not stem the tide of the increasing vertical (and horizontal) integration of kids' media. On the contrary, media companies became even more creative in their strategies to produce strong brands. One of these strategies involved the cable industry. Cable television has much more effectively captured a child audience (even while it may be smaller than broadcast networks) through its ability to produce dedicated children's channels, such as Nickelodeon and the Disney channel, as discussed in the next section. In part because of the success of these kinds of channels, the strategy of broadcast television has fundamentally changed so that, in the current environment, most broadcast children's programs are, in fact, purchased from cable stations. For example, CBS (along with Nickelodeon, a division of Viacom) airs Nickelodeon programs on Saturday mornings and ABC (which is owned by Disney) features Disney-branded programs, etc.[13] Efforts to create and establish children's programs on network television are now generally understood as too financially risky; more children watch prime-time television on the networks than programs that are designed specifically for children, which makes the lucrative prime-time period a better use of a network's development budget.

The Cable Industry: The First Stage of Post-Network Television

The advent of the cable industry in the late 1970s and the creation of dedicated kids' channels signaled a crucial change in the youth television landscape. During this period, satellite technologies and an increasingly relaxed regulatory environment provided the technological, cultural, and economic context for cable technologies to emerge as a specific and discrete industry in the United States. The installation of cable technology in the American home during the 1970s was positioned and celebrated in terms of its difference from broadcast television. Cable television could offer less "lowest-common-denominator" and crassly commercialized television, less intrusive advertising, more interactivity on the part of the viewer, and generally more viewer empowerment. In other words, cable television would ostensibly "serve" the public interest in a way that broadcast television ignored.[14]

Indeed, because cable television does not use publicly owned airwaves, it is not subject to the same public interest obligations as broadcast television. Despite this, however, the optimistic rhetoric structuring the early cable industry implied that cable could address some of the problems broadcast television had in fulfilling the public interest obligations of the media. The cable industry was flexible in ways that broadcast television was not because cable is not as dependent on advertising revenue. Additionally, cable channels could afford to experiment with different formats and content in the early years of the industry because there was no clear recipe for success for this technology. Cable seemed poised to provide access to a greater variety of media forms and points of view than could be found on broadcasting sources, as increasing channel capacity, a pared-down regulatory apparatus amenable to growth of the industry, and consumer demand for new services transformed how homes received television signals. The perception of cable, within the industry and among the public alike, was that this new industry was an important alternative to the broadcast monolith.[15] Because cable television provided an alternative to broadcast television, the emergence of this industry fundamentally transformed the children's television landscape. The opportunity to take risks in the new industry meant that channels could experiment with the broadcast format for children's television, leading eventually to dedicated channels for children such as Nickelodeon and the Disney channel. These cable channels were able to make use of a broader range of time slots and could create original programming (and thus were less beholden to toy companies for licensed character-driven programs), resulting in programming that looked different from the typical broadcast fare. Channels such as Nickelodeon were able to craft a unique television identity within a competitive field because of the (initial) flexibility of the cable industry.

Nickelodeon is a very successful cable channel and has been noted frequently as a trailblazer in the realm of children's television in general. A division of Viacom International, Inc., and delivered via satellite, the 24-hour Nickelodeon channel is an outlet of children's programs with millions of subscribers that captures enormous children and adult audiences. It has been recognized by both industry professionals and media scholars as one of the most successful innovations in cable programming. In part, this is because of its segmented programming that appeals to all age groups: pre-school programming in the morning (the Nick Jr. programming block, which includes *Dora the Explorer* and *Go Diego Go!*), young children's programming in the afternoon (for instance, the popular *The Fairly Odd Parents* and *SpongeBob SquarePants*), pre-adolescent and adolescent programming in the later afternoon (*Drake and Josh*, *iCarly*), and "tween" programming in the evening (teen variety shows such as *The Kids Choice Awards*, *Zoey 101*, and MTV-style variety programs that encourage audience participation). Then, in the later evening, Nickelodeon airs *Nick at Nite*, which is a nostalgic programming line-up of older family shows ranging from *Petticoat Junction* to *Home Improvement*. Aside from the actual program line-up, Nickelodeon has also garnered critical acclaim for its original programming featuring both animated and non-animated programs that address "kids as kids."[16]

Nickelodeon is not simply a random success story in the often-unpredictable world of television, however. The dedicated children's channel is a direct result of the influx of cable television in the United States, an influx that has since become normalized. Indeed, by mid-year 2005, 94.2 million households—86 percent of households with televisions—subscribed to some sort of multichannel video programming distribution.[17] The cable television landscape involves a proliferation of formats and channels and a disaggregation of a once-mass television audience. The efforts spent by advertisers to determine the broadest possible audience in the early years of broadcast television are now streamlined into niche markets, where one can tune into highly specific channels, from golf to music to children's programming.

The advent of cable television was a significant intervention in children's television practices, especially in the way that cable channels such as Nickelodeon and Disney have captured the children's audience through age segmented programming and ancillary products. As I explore in *Kids Rule! Nickelodeon and Consumer Citizenship*, Nickelodeon's original media productions represent a shift not only in the kinds of shows that appeal to children, but also in how children are addressed in and through the media. By marketing itself as a "kids-only zone," Nickelodeon clearly encourages its audience members to understand themselves as different from adults, creating not only a divisive line between the cultural realms of children and adults, but also an opportunity for children in the audience (or adults participating as members of a child audience) actively to construct themselves as particular kinds of citizens.[18] At the same time, the network is a dominant force in the construction of children's consumer culture, and unmistakably addresses its youth audience as consumers through its programming, advertising, and products. This tension between consumer citizenship and political citizenship is necessarily maintained by the political economy within which Nickelodeon is situated; it is precisely the structure of the cable industry that allows for a more innovative and active address to a youth audience even as this industry remains structured around competitive commercial interests. Because of the flexibility of a cable network's programming—where channels such as Disney and Nickelodeon air exclusively children's programs—there are more resources, and thus more potential, to create innovative and even at times risky programs (especially when compared with the more economically constrained broadcast networks). As recent technological innovations in animation as well as a broad diversifying of content in children's programs have demonstrated, the children's cable industry is clearly a valuable cultural site in which to locate recent shifts in the structure of kids' television.

Cable channels such as Nickelodeon, Disney, Fox Kids, and Cartoon Network have clearly learned the lesson that earning the early loyalty of a child audience is an important marketing strategy, given that children represent not only a primary market, but also a future and influence market. Combined with the continued reach of the cable industry into more and more niche channels such as those for children, the brand identity of channels and specifically the marketing of brands as particular experiences have achieved a new economic significance. It is to this brand context that we now turn, to examine what might be called the second stage of "post-network" television for children's television.

Brand Culture: The Second Stage of Post-Network Television

In the July 24, 2008, issue of *KidScreen Daily*, a trade e-magazine for children's media, there was an announcement of a new partnership between Wal-Mart and

Disney themed around Disney's hit kid program *Hannah Montana*. The new partnership kicked off with a back-to-school marketing strategy, inaugurated by a service offering free wake-up calls and customized activity reminders from none other than Hannah Montana herself. By simply logging onto a website, parents could select a date, time, and theme for their child's wake-up call, and choose between three different messages.[19]

In the June 25, 2008, issue of *casualgaming.biz*, there was another announcement, this time by Nickelodeon and their games division, that the network (itself part of MTV networks, which in turn is part of Viacom) planned to explore new areas to develop casual game markets (casual games are those typically played on a personal computer online in web browsers). As Dave Williams, head of the Nickelodeon games division, commented: "it is time for us in the casual games business to challenge the traditional notion of who we are building games for and targeting their needs."[20] The article continues:

> "Go beyond the Soccer moms," he said, saying that building online web games with just the stereotypical 35+ female gamer in mind has "become self-limiting … It's vital we look beyond our audience for the traditional industry to grow." He elaborated: "Moms, dads, teens are all playing casual games in droves. This shows there are huge untapped markets in this space." Nickelodeon has identified key audience segments, he explained, [which] include "Time Fillers," who account for 17 per cent of the audience and play online games just for fun; the "Gaming Enthusiasts," who are 19 per cent of the market and are challenge-loving players; "Guilty pleasurists," another 17 per cent chunk who play games to blow off steam; and "Average Joes," the 19 per cent of the audience who see games as a form of social currency—but don't tell people about it.[21]

The levels of complexity in these marketing strategies (segmenting audiences narrowly enough to include the range of "time fillers" to "average Joes," for example) and the personalization of the audience (Hannah Montana waking up one's child with a phone call and a personalized message) is indicative of the scope and reach of media brand culture for kids in the contemporary environment. To bring us back to the anecdote with which I began this essay, for kids, "watching" television increasingly means immersing oneself in a branded world of multi-media texts (accompanied by multi-tasking within those texts) and interactive technologies. While television remains a central part of the kids' media environment, it is crucially interconnected with other media forms, including not only interactive web-based activity and video technologies, but also a general immersion within brand culture. To be a "fan" of, say, the television program *Hannah Montana* on the Disney channel means not only watching the show, but also going to concerts (and sometimes paying $2,000 for tickets!), watching streaming videos on a website, purchasing CDs, playing Hannah Montana-themed video games—and being awakened for school by Hannah Montana, with a cheery "It's time to wake up and get a move on!"[22]

What does the increasing ubiquity of brand culture and interactive new technologies mean for kids' television viewing practices, and for kids' television itself? The complex marketing strategies structuring children's media culture belie a simple goal: gain brand loyalty as early as you can. According to marketing expert James McNeal, the kids' market has been growing at a rate of 10 to 20 percent a year since the mid-1980s. In 2001, four- to twelve-year-olds had their own annual income of approximately $40 billion, and children as an influence market were responsible for over $300 billion.[23] By 2004, McNeal estimates that children directly influenced $330 billion of adult purchasing and "evoked" another $340 billion.[24] This number continues to grow, offering clear evidence that kids are at the center of consumer culture. As Juliet Schor argues:

> the most important change in [contemporary] consumer culture was not what the analysts were focusing on—Internet shopping, branding, consumer credit, or customization of products. It was that the imperative to target kids was remaking the marketplace. By 2003, Martin Lindstrom, one of the world's leading branding gurus, opined that 80 percent of all global brands now required a tween strategy.[25]

Not surprisingly, the early loyalty of a child audience is also important for television networks, and, while I agree with Schor that kids are a fundamental element in the changing media marketplace, I also think that the other elements of this change, such as branding and customization of products, are crucial to take into account when considering exactly *who* the "kid" is in the contemporary media environment. In other words, thinking about consumer culture at the level of literal purchasing does not help in understanding the relation between kids and media—or more specifically, in the relationship of kids with television in a post-network era. When attempting to understand this relationship, it is necessary to shift focus and to move our gaze from the relatively insular relationship between a child viewer and a television program as a way to find out what kind of "effect" the media has on kids, or what kind of television "fan" a child might be, to the more complex and multi-dimensional relationship of a child within media culture. It also requires a different look at kids' television programming, as not only are shows connected to ancillary markets and practices through websites and user-generated content, but these technologies increasingly hold a prominent place within the actual program narratives.

Within this kind of economic environment, where kids are thought to represent "more market potential than any other demographic group," competition for the attention and loyalty of children within media companies has become even more intense. Not surprisingly, children themselves have a new economic significance as consumer citizens. Consumer citizenship indicates a certain willingness to participate in consumer culture through the purchase of goods as well as a more general affirmation of consumption habits, but also points to something broader, where the distinctions between cultural and social practice and consumption are not so finely drawn. The contemporary consumer citizen is situated within the context of brand culture, and any quick glance at twenty-first-century cultural, social, and political life in the United States discloses compelling evidence that, regardless of gender, ethnicity, or socio-economic status, this is where we live our lives. While advertising continues to have a dominant presence in both public and private spaces, what characterizes contemporary culture is not so much the ubiquitous advertisement, but rather the normalization of brand culture—broadly defined as the deliberate association of a product with an idea or a concept as well as a trademarked name. Within this context, consumer participation is indicated not simply (or even most importantly) by purchases made, but by brand loyalty and affiliation, a connection that links brands to lifestyles, to politics, and even to social activism. Brand culture thus shapes not only consumer habits, but also all forms of political, social, and civic participation.

Effective branding strategies that result in attracting both narrowly specific audiences and advertisers concerned with reaching those same specific audiences have become the norm for transnational media conglomerates such as Viacom.[26] Nickelodeon, Disney, Fox Kids, and Cartoon Network have all assiduously crafted their brand identity so that kids watching these networks affiliate more with the brand than with individual programming. (Of course, there are exceptions to this, as with the *Hannah Montana* phenomenon. But even that particular program is much more than just a television show, with music, film, and ancillary products bringing in more revenue than the actual television program.) The far-reaching branding strategies of children's television channels reveal an important shift in the post-network era; a brand identity requires being more than just a television show, and thus it requires a strategy that is often not available to a "broad"caster. The shift from program loyalty to brand loyalty reflects another, broader cultural shift, in which children are understood (and marketed to) not only as savvy consumers who make choices in the marketplace independent of their parents but also as sophisticated practitioners of "convergence culture."[27] In other words, the "social practice" of television watching, in the contemporary context, is as much (if not more) about children engaging in multi-media activities and interacting with brand culture as it is about actual television viewing. This cultural shift in youth subjectivity is reflected in the thematics of television programming itself.

Consider, for example, Disney's program *Hannah Montana*. The show revolves around a young teenage girl, named Miley (played by Miley Cyrus), who has an alternative identity as a rock star, Hannah Montana. The show switches back and forth between the typical quotidian routines of a tween in contemporary culture when it centers on Miley, and the glamorous life of a rock star complete with sold-out concerts and musical routines when the lens shifts to Hannah. The show thus celebrates the intersections of commercial pop culture with everyday teenage life, and validates—to say the least—the practice of multi-tasking within and among one's different personal identities. In an even clearer example, the Nickelodeon television program *iCarly* incorporates brand culture and multi-media activities as the primary topic of the show. *iCarly* revolves around three teenagers, Carly, Sam, and Freddie (also featuring Carly's older brother Spencer), who produce their own program as a webcast. The webcast (called "iCarly") not only presents material from the Nickelodeon writers themselves, but also solicits "real" viewers to send in content to use in the program's webcast. As Ethan Thompson has argued,

> That's a predictable enough gimmick for a TV show about a webcast, but it is also a manifestation of how Nick has connected with a generation of viewers for whom it's perfectly natural to watch clips of a TV show online as well as have nonlinear editing software installed on their home computer.[28]

Nickelodeon has been particularly successful at this kind of "connection with a generation of viewers," and the incredibly popular *iCarly* (the debut episode in 2007 attracted 3.5 million viewers, and the numbers have increased since then) is a telling example of how children's television incorporates a broader, multi-media format as an integral theme in programming. Both *Hannah Montana* and *iCarly* are heavily steeped in brand culture, from clothing to food products to computer technologies (*iCarly* is particularly clever with the integration of a brand aesthetic, featuring, for example, laptop computers with a pear shape icon rather than the apple, clearly signaling Apple computers but without direct product placement).

The narratives of kids' programs like these make it clear that early models of examining what kind of "effect" the media has on kids, or what kind of television "fan" a child might be, are no longer sufficient to understand the position of the child viewer immersed in the more complex and multi-dimensional media environment. Rather, the narratives of kids' television incorporate the shifted conception of the child her/himself, as an individual in relation to brand culture and multi-media environments. To be sure, networks such as ABC, CBS, and NBC continue to produce (or buy from syndicates) children's programming, but the focus of the programming and the advertisers is even more on the ancillary products and activities of the actual program itself—sell-through videos, manufacturing, toys, and websites.[29]

Kids in Brand Culture: Victims or Savvy Consumers?

Understanding children's television, then, in a post-network brand culture means that we need to look not only beyond actual television programming but also beyond the television industry in order to understand what it means to be a child "fan." Liz Moor, in her book *The Rise of Brands*, argues that brand culture needs to be understood as a particular kind of media space, one that traverses beyond the traditional venues for advertising and promotion of goods. As she argues,

> Those working in branding tend to take it for granted that "everything is media"; that any site where a brand appears is a potentially communicative medium. "Media" in this sense, are not a discrete entity or set of entities; they are simply the context in which all marketing takes place.[30]

Again, this is not to say that the numerical figures concerning how many hours American children are sitting in front of a television set should be disregarded, but simply to argue that, in the present cultural economy, the relationship of children to television goes beyond actual viewing practices. Understanding children's television in a post-network society means considering the broader media space—the space of brand culture—that constructs the relationship children have with television as well as other media artifacts. As Moor continues,

> Branding in this sense is a kind of spatial extension and combination, in which previously discrete spaces of the brand—the advert, the point of purchase, the product in the home—are both multiplied, so that there are simply more "brand spaces," and made to refer back and forth to one another so that they begin to connect up or overlap.[31]

This overlapping, ever-increasing space of brand culture has led to a predicable analytic binary when it comes to examining children and media. The early adoption of television into American homes witnessed cultural debates over whether the child's relationship with television should be considered as one in which the child is a victim or a beneficiary. Deregulation in the 1980s and the emergence of the cable industry reinvigorated these debates, focusing on violence, sex, and commercialism and the impact these themes might have on child viewers. These debates persist regarding twenty-first-century brand culture and interactive media technologies, where public discourse continues to frame kids who participate in and with the media as either impressionable victims or savvy entrepreneurs. For example, in the last decade there has been a heightened debate about the consumption habits of young American kids, articulated in popular books that lament their spending habits, such as Schor's *Born to Buy* and Alicia Quart's *Branded.*[32] News magazines moralize about superficial spending, telling parents to "just say no" to their kids' extravagant demands, and there are multiple themes in children's television shows about the problems that come with being an excessive consumer at so young an age. Not surprisingly, alongside these popular laments about the commercial corruption of America's youth there is another, more optimistic interpretation of the current generation's spending habits. This interpretation is primarily an economic one, stemming from marketers, that celebrates the child consumer. And, as I've stated, in the early twenty-first century, children *are* spending more and more; they have increasingly sophisticated consumption habits, they continue to influence the purchases their parents make, and their general culture is a branded one. Marketers and branders have been very quick to pick up on kids' consumer culture, and continue to strive to create a branded ethos in which children organize not just their spending habits, but their relationships, affiliations, and lives.

The obvious contradictions that characterize these debates mirror some of the contradictions surrounding how young Americans are interpellated as both citizens-in-training and consumers in contemporary culture. Indeed, there are very real and direct parallels between the self-construction and interpellation of children as consumer citizens and their consumption habits. The differences in these two responses also offer the opportunity to parse the contradictions within media texts and consumer reception in the contemporary context. In other words, the child consumer citizen exists neither utopically nor dystopically within brand culture and the new media environment. Yet, we seem to linger in an either/or paradigm, where kids are understood either as impressionable victims of corporate media or as sophisticated consumers.

For example, the brand context for contemporary children's television shows has signaled an alarm for some, who argue that the increasing number of advertisements available to children on a growing number of platforms means that children are encouraged to be better consumers at an even younger age. And it certainly is true that marketers have quickly picked up on the sophistication of children using interactive media such as the Internet. As Dale Kunkel argues, "marketers may use interactivity to target particular products or ad strategies to individual viewers in a manner that significantly increases their persuasive power."[33] As Kunkel points out, while there are current FCC policies that limit the amount of time advertisements are allowed within children's programming, "interactivity would potentially allow children to spend unlimited time in highly commercialized environments with a quick click of the remote control."[34] If being "interactive" means simply to be a more savvy consumer and to be able to navigate between a program aired on television, a website, and a video game, then it does indeed seem as if the doomsayers that lament the fate of the "branded child" are on to something: *iCarly* could be considered perhaps one of the most sophisticated of these kinds of gimmicks. However, the idea that interactivity means spending "unlimited time in highly commercialized environments" assumes that there is space *outside* these commercialized environments, a space for perhaps a non-commercial kind of interactivity to take place. Clearly, interactivity in the contemporary kids' media environment is connected to brand culture, but calls for a more sophisticated analysis because the relationship between the child viewer and media production is less controlled and perhaps more unpredictable in the current environment.

This doesn't mean that television broadcasters are rendered obsolete, but it does mean that these broadcasters will have to tap into a multi-media environment in order to maintain their brand. As Stephen Rockwell points out,

> With a superabundance of choice, creative marketing of products will become increasingly important to drive content consumption. Whether this is accomplished via direct marketing or through techniques such as promotional tie-ins, content providers with larger budgets will create more demand for their products and likely make it difficult for smaller companies to compete.[35]

Thus, brands such a PBS Kids, Nickelodeon, and Disney remain big players in the children's television landscape, but the tried-and-true forms of marketing to children are no longer sustainable in an interactive environment.

Another factor that must be taken into account within contemporary media culture is the rise of user-generated content, where kids participate in the development of a brand through online competitions, creating videos, and advertising for television and other media on personal webpages on social networking sites. The increase in this kind of activity cannot be ignored: social media usage grew 668 percent within the period of April 2005 to April 2007.[36] While this includes adults, it is clear from the normalization of social networking sites such as MySpace and Facebook, video sites such as YouTube, and the success of programs such as *iCarly* that children and teenagers are part of this statistic. Rather than treating consumers as a large (even if narrowly construed) demographic, branding agents and marketers have shifted their strategies to treat consumers as individuals: "treating customers as individuals requires more than a technology strategy. For most brands, it requires a new kind of business strategy born of a fundamental shift in thinking."[37] This "shift in thinking" means that, from at least one important perspective, in the Web 2.0 world of branding, children can be exploited as consumers in an ever-more insidious way, where every social act can be parlayed into a profitable function for media companies. Within this perspective, the labor of children is directed toward further developing of the brand, whether in terms of actually creating ads through competitions or simply promoting brands on personal webpages. Again, while this may be true in particular instances, this analysis also misses how user-generated content is used as critiques of media as well, challenging the domination of media industries over individual users.

Television in the Space of the Brand

The question, then, that emerges from the contemporary kids' media and televisual context involves the impact of this new environment: is profit for media corporations the only effect we witness? Are there social shifts or changes in individual identities that also occur within brand culture that may or may not indicate profit motive? How is the "social practice" of watching television a different kind of practice in post-network brand culture? Aside from the commercially driven interpretation of children as consumer citizens, there is yet another optimistic voice in the fray, emerging from those who study new media technologies and the current generation's skill and acuity at navigating in a Web 2.0 world. Indeed, while "interactive" certainly indicates a particular engagement with commercial messages on multi-media platforms, it is also not the only definition of the term. Part of the difficulty in figuring out what children's television truly *is* in the contemporary era means that we need a sharply honed sense of the contradictions involved in both cultural production and cultural consumption, as well as within the ways that the subject position of consumer and producer continually merge. Children can no longer be considered simply as an "audience" of television, but rather as part of the production of a broader television environment. As television continues its trajectory of being just one element in an increasingly interactive and sophisticated technological environment, the concept of the "youth audience" needs to be broadened as well.

For instance, in a recent *New York Times* article, "Sorry, Boys, This Is Our Domain," the reporter states that "the cyberpioneers of the moment are digitally effusive teenage girls." Regardless of the questionable gendered assumptions of this article (most of the blogs produced by girls focused on fashion, cooking, or gossip, and the *New York Times* published the piece in its "Style and Fashion" rather than its "Technology" section), it did state some important statistics, such as a recent Pew Internet and American Life Project study that found that 70 percent of girls aged fifteen to seventeen and 57 percent of boys of the same age have either built or worked on websites and have created profiles on social networking sites.[38] The Internet is a highly commercial realm—but it is also difficult to regulate and control. Children and youth are, in fact, "pioneers" of cyberspace, and the activity occurring in this cultural realm cannot be mapped on to a traditional linear model of messages and reception, such as that which has been used to measure the impact of television on children. How, for example, do we think about the audience for *iCarly*, when some kids in that audience produce home videos that end up being part of the fictional webcast featured on the program? Or how do we think of Hannah Montana fans who not only watch the program, but also buy tickets to see a rock concert performed by an actress who plays a rock star on a TV show? With kids' increasing access to new technologies (ranging from social networking sites to TiVo), ever-growing niche markets, and interactive television shows, efforts to predict the behavior of the youth television "audience" are increasingly more complex. Unlike the earlier era of kids' television, where there was tighter control over the messages and images that were created, the current environment is more indeterminate, allowing for the potential (even if rarely realized) of a shifted kind of youth agency.

Brand culture functions as a kind of lifestyle politics for kids—something one is, or does, rather than pointing to a particular consumer good one purchases. How is attending to the relationship between kids and brand culture a means to conceptualize the ways in which youth culture is situated in the contemporary media economy? The "social practice" of watching television is clearly a dynamic one, adapting and moving with other changes in the media environment. When examining children's television, it makes sense to critically interrogate the concepts we use to determine media subjects engaged in the practice of watching television, such as "consumer" and "producer." What is at stake is not simply revisiting these terms to theorize what place they might hold in a cultural debate about the making of identity. Rather, a new conceptualization of these terms and the contradictions between them is needed as a way to account for changing practices of cultural production within a shifting televisual economy, and will allow us to have a much more engaged sense of what the relationship between youth culture and "television" is about. In order to analyze the "practice" of watching television in the contemporary cultural and technological economy, we need to resituate how we think of children—and childhood itself. There is much discussion about new media technologies for children, new media platforms, new ways for marketers to target kids. There is less discussion about how these new contexts become the ground in and through which children craft identities and subjectivities. If television watching is indeed a practice, one that involves a variety of elements involved in interplay (such as medium, viewer, production, cultural economy, etc.), then the traditional ways of understanding children as media watchers—as in either the victimized or savvy consumer model—no longer suffice. When I asked my daughter why she prefers watching "television" on the computer, her response was given in a rather perplexed fashion, as if the answer should be obvious to me. Despite the fact that this way of watching television is normative for her, the answer *isn't* obvious. In the contemporary brand environment, where media is shaped within a multi-dimensional framework, where producers are often consumers, and consumers are just as often producers, the "social practice" of children watching television involves a series of complex activities, the "effects" of which require a similarly complex understanding of both the child and television itself.

Notes

I want to thank Amanda Lotz and Cara Wallis for their insightful comments and suggestions for this essay.

1 David Morley, *Family Television: Cultural Power and Domestic Leisure* (London: Routledge, 1988); Stuart Hall, "Encoding/Decoding," in *The Media Studies Reader*, ed. Paul Harris and Sue Thornham (New York: New York University Press, 2000).

2 Lynn Spigel, *Make Room for TV: Television and the Family Ideal in Postwar America* (Chicago: University of Chicago Press, 1992); Heather Hendershot, *Saturday Morning Censors: Television Regulation before the V-Chip* (Durham, NC: Duke University Press, 1998); and William Boddy, *Fifties Television: The Industry and its Critics* (Urbana: University of Illinois Press, 1990).
3 Donna Mitroff and Rebecca Herr Stephenson, "The Television Tug-of-War: A Brief History of Children's Television Programming in the United States," in *The Children's Television Community*, ed. J. Alison Bryant (Mahwah, NJ: Lawrence Erlbaum Associates, 2007), 10.
4 Spigel, *Make Room for TV*, 2–3.
5 Mitroff and Herr Stephenson, "Television Tug-of-War."
6 Spigel, *Make Room for TV*; Hendershot, *Saturday Morning Censors*; Boddy, *Fifties Television*.
7 Cited in Spigel, ibid.
8 Jason Mittell, "The Great Saturday Morning Exile," in *Prime Time Animation: Television Animation and American Culture*, ed. Carol Stabile and Mark Harrison (London: Routledge, 2003), 13. See also Mitroff and Herr Stephenson, "Television Tug-of-War."
9 Mittell, ibid., 50.
10 For more on this, see Tom Engelhardt, "The Strawberry Shortcake Strategy," in *Watching Television*, ed. Todd Gitlin (New York: Pantheon Books, 1986), 68–110; and Robert Horwitz, *The Irony of Regulatory Reform: The Deregulation of American Telecommunications* (New York: Oxford University Press, 1991).
11 Joy Tashjian and Jamie Campbell Naidoo, "Licensing and Merchandising in Children's Television and Media," in *The Children's Television Community*, ed. Bryant, 165.
12 Amy Jordan, Kelly L. Schmitt, and Emory H. Woodard, IV, "Developmental Implications of Commercial Broadcasters' Educational Offerings," *Journal of Applied Developmental Psychology*, 22, 1 (2001), 87–101.
13 Terry Kalagian, "Programming Children's Television: The Cable Model," in *The Children's Television Community*, ed. Bryant.
14 Sarah Banet-Weiser, Cynthia Chris, and Anthony Frietas, eds, *Cable Visions: Television Beyond Broadcasting* (New York: New York University Press, 2007).
15 Tom Streeter, "Blue Skies and Strange Bedfellows: The Discourse of Cable Television," in *The Revolution Wasn't Televised: Sixties Television and Social Conflict*, ed. Lynn Spigel and Michael Curtin (New York: Routledge, 1997); and Megan Mullen, *The Rise of Cable Programming in the United States: Revolution or Evolution?* (Austin: University of Texas Press, 2003).
16 Sarah Banet-Weiser, *Kids Rule! Nickelodeon and Consumer Citizenship* (Durham, NC: Duke University Press, 2007).
17 Banet-Weiser et al., *Cable Visions*, 2.
18 Banet-Weiser, *Kids Rule!*
19 "Hello, Hannah, Is it Me You're Looking For?," *KidsScreen Daily*, July 25, 2008.
20 Michael French, "Nickelodeon Predicts 'New Wave' for Online Casual Titles," accessed on *casualgaming.biz*, July 24, 2008.
21 Ibid.
22 www.hannahmontanacalls.com, accessed July 25, 2008.
23 James McNeal, *Kids As Customers: A Handbook of Marketing to Children* (New York: Lexington Books, 1992).
24 McNeal, cited in Juliet Schor, *Born to Buy: The Commercialized Child and the New Consumer Culture* (New York: Scribner, 2004), 23.
25 Schor, ibid., 12.
26 Elizabeth Hall Preston and Cindy L. White, "Commodifying Kids: Branded Identities and the Selling of Adspace on Kids Networks," *Communication Quarterly* 52, 2 (spring 2004), 115–28.
27 For more on this, see Henry Jenkins, *Convergence Culture: Where Old and New Media Collide* (New York: New York University Press, 2006).
28 Ethan Thompson, "I Want my TweenTV: iCarly, Sitcom 2.0," http://flowtv.org/?p+1597, accessed August 10, 2008.

29 Of course, ancillary products have long been important to children's television, as we see with toy tie-ins and the cultivation of particular child stars, such as the Olsen twins, as a kind of media "brand." But in the contemporary environment, the products associated with television shows often overshadow the program itself. This, combined with interactive technologies and a focus on user-generated content, forces us to shift our focus from the program to the multitude of cultural products that are in some way connected with television.

30 Liz Moor, *The Rise of Brands* (Oxford: Berg Publishers, 2007), 26.

31 Ibid., 47.

32 Schor, *Born to Buy*; Alicia Quart, *Branded: The Buying and Selling of American Teenagers* (New York: Perseus, 2003).

33 Dale Kunkel, "Kids' Media Policy Goes Digital: Current Developments in Children's Television Regulation," in *The Children's Television Community*, ed. Bryant, 219.

34 Ibid.

35 Stephen Rockwell, "Networked Kids: The Digital Future of Children's Video Distribution," in *The Children's Television Community*, ed. Bryant, 193.

36 Kelly Mooney and Nita Rollins, *The Open Brand: When Push Comes to Pull in a Web-Made World* (Berkeley, CA: Aiga Design Press, 2008), 79.

37 Ibid., 123.

38 Stephanie Rosenbloom, "Sorry, Boys, This Is Our Domain," *New York Times*, February 21, 2008.

Chapter 5

National Nightly News in the On-Demand Era

Amanda D. Lotz

The death knell for the nightly network newscast has been ringing for over twenty years now. Even seventeen years ago, in 1992, CBS Washington bureau chief Barbara Cohen noted, "It has become fashionable to predict the demise of network news in general and the evening news broadcasts in particular," as she then added to the chorus she described while making some exceptions.[1] Perhaps this phenomenon can be dated to an even earlier moment, but refrains forecasting the end began with considerable regularity by 1986, the year Fred W. Friendly, former president of CBS News and professor emeritus at Columbia's Graduate School of Journalism, publicly opined that, "Unless the networks make their product appreciably and dramatically superior, I doubt there's much of a future for network news."[2] More than two decades later these newscasts remain on daily schedules, and arguably not as a result of the appreciable improvement he suggested. Since then, the newscasts, the networks, and even television-at-large has come to be challenged in ways these commentators could not have imagined.

Nevertheless, Friendly delivered his comments at an important moment for network news; by 1986, all three networks were in the midst of ownership changes that would bear significant consequences for their news divisions and were engaged in often public battles over resulting budget cuts. Also by 1986, many of the first inklings of the multi-channel transition were beginning to affect network operations. Viewer adoption of technologies such as VCRs and remote controls might not have influenced news as significantly as other program forms, but cable competition was beginning to concern network news departments as CNN and Headline News appeared to be more than passing phenomena.

The nightly network newscasts, by which I mean the half-hour nightly news produced by ABC, CBS, and NBC and carried by most local affiliates at 6:30 each evening (5:30 Central),[3] once defined network television news. In their much storied past, the three broadcast networks drew 72 percent of those watching television at 6:30 to these newscasts; but television news and television in general have changed substantially since that era.[4] By 1999, the three national broadcast newscasts gathered just 47 percent of the audience, and viewership dipped to 37 percent by 2005.[5] Surveys by the Pew Research Center for People and the Press in 2006 stated that only 28 percent of those surveyed reported regularly watching nightly network news, a figure that had dropped from 60 percent in 1993, and indicated that the number of viewers watching network newscasts was smaller than those viewing local (54 percent) or cable news (34 percent).[6] Comparisons between network and cable news are complicated, however, by the different audience counting methods that result from their varied formats. Although 34 percent of viewers may watch some cable news, there are rarely more than a million or so watching a cable newscast at any particular moment. Nearly ten times more people watch the nightly network newscasts than prime-time cable news shows, yet so many viewers report receiving news from cable because it is always available.[7]

But, regardless of audience erosion, network newscasts remain an important part of the broadcast television day, and the 28 percent of those who remain—an estimated 26.1 million—encompass a sizable collection of viewers in today's narrowcast media environment.[8] Peggy Green, president of broadcast at media buyer Zenith Media, is on the mark in noting that, for advertisers, news is "a vanity daypart more than anything else," as advertisers seek for the cultural capital of news to be associated with their product.[9] Historically, news has been a vanity point for the networks as well—as news has always cost more than it returned in revenue, but brought prestige to the network and the suggestion of public service. The newscasts offer some financial value to the networks, and each still contributes roughly $150 million in annual advertising revenue.[10] *Wall Street Journal* reporter Kyle Pope noted that the costs of newsgathering typically outpace these revenues considerably, and estimated the costs of a single network news division at $400 million per year in 1999, making profit unlikely.[11]

Despite the continued erosion of audience share, mounting financial losses for the newscasts (as well as the news divisions and the networks in many cases), and the reality that the newscasts and their cultural capital have been significantly reduced from their former identity, network news continues to be relevant to the future of television. This chapter explores how national network news is changing in response to competitive and technological developments arising throughout the multi-channel transition and post-network era to examine both challenges and possibilities now confronting news. In the post-network era, online and cable competitors able to provide live coverage of events throughout the day threaten the relevance of evening network newscasts, yet significant prospects also exist in the new media space for the networks to re-establish their role in daily news provision. The convention of a daily time-prescribed newscast that is part of a linear schedule may be in peril; however, many of the technical and industrial challenges that have long limited network news also may be assuaged by developments characteristic of the post-network era.

Television news continues to be an important site of study for understanding how this still powerful form informs societies and enables and hinders the proper functioning of democracy. The substantial changes in the industry's operation and viewers' use of the medium, however, require us to revisit some of our central assumptions about television news and divorce our hypotheses about its future from the expectations held for prime time. The consideration here focuses on only one aspect of television news—that of the national network nightly newscast; the situation of local news production in the post-network era, while as important, has too many other dimensions to allow joint assessment and is addressed in chapter 8.

Network News During the Network Era and Multi-Channel Transition

As suggested by quantifiable factors such as audience size and revenue figures, network-era television news derived cultural importance from its vast reach and ability to provide a narrative about contemporary events for much of the sizable television audience. The first television newscasts evolved from radio, where news developed relatively late in comparison with other programming forms and did not become a major component of programming until World War II loomed.[12]

Television newscasts date to the medium's early days, although it took some time to establish the format and features of the newscast common throughout much of the network era. Despite technological limitations now taken for granted—such as the previous impossibility of live video from around the world—the television news divisions provided many Americans with a window on national and international events. Network news divisions delivered many of the key moments in television history—such as the moon landing and multiple days of coverage following the Kennedy assassination. The video images and the immediacy with which viewers saw such events contributed to the particular place and role in their remembrance afforded to television. Throughout the network era, network news generally provided a respected and trusted source of information.

The general business of television news changed little until the mid-1980s. Newscasts expanded from 15 to 30 minutes in 1963, and news reporting was advanced significantly with videotape and satellite links, but the programming form otherwise maintained considerable constancy. News divisions also produced the highly regarded television documentaries of the 1960s that aired during prime time and indicated some of the greatest potential of television journalism.[13] While the documentaries occupy only a brief moment of network history, the news divisions re-emerged in prime time in the late 1970s with news magazines such as *60 Minutes* and *20/20* that offered substantial financial value. They deliver high ratings among advertiser-coveted viewers and require minimal production costs in comparison with prime-time narrative series.

Like other program forms, news operations began experiencing major adjustments in the mid-1980s—although network newsrooms were affected by different aspects of the multi-channel transition than those most significant for other parts of the program day. Devices such as the remote control and VCR introduced minimal adjustments because of the particular significance of immediacy as a program quality for news. Even the arrival of the broadcast network FOX in 1986 brought little change, as the network did not include a national newscast in its program offerings; nor did broadcast networks the WB and UPN a decade later.

Rather, the changing corporate ownership and management of each of the Big Three networks in the mid-1980s produced some of the most significant changes for network news divisions. As the networks became cogs in diversified corporate portfolios, the high cost of their news divisions came under assault and the latter endured massive budget and personnel cuts. CBS laid off nearly 500 news staff members and was forced to reconfigure its operations with $30 million less in its news budget in the mid-1980s as a result of new owner Lawrence Tisch's demands of profitability.[14] NBC had actually enacted similar cuts a decade earlier as then owner RCA shifted its innovation emphasis away from the seemingly mature business of television.[15] Nonetheless, the news division was expected to cut 5 percent of its costs alongside the rest of the network as in the mid-1980s new owner General Electric sought to apply its tight fiscal management techniques.

For each network, and particularly their news divisions, this new ownership introduced vast changes to network culture as the new corporate owners did not regard the public trust function of their acquisitions with much seriousness and concerned themselves foremost with cost cutting. Although the networks had developed many layers of management that could do with trimming and had grown accustomed to some unnecessary excess, the fervor of the fiscal management of the new owners baffled those at the top of the news ranks. Corporate executives such as GE's Jack Welch for NBC and Lawrence Tisch at CBS sought answers to questions such as "What is the average cost of a story?," which news division heads had never considered, let alone compiled.[16]

Some might argue that it is difficult to say the networks regarded serving the public trust all that seriously when they were independently owned, but in that era the networks often turned to their news coverage to justify broadcast licenses that were effectively a subsidy worth millions and eventually billions of dollars.[17] Although the vaguely defined requirements of public service are mandated to the stations that hold the broadcast licenses rather than the networks, throughout much of television history the networks operated their news divisions with an acknowledgment that this programming was distinctive from the rest of their schedules and was a crucial service to their affiliates. Each network has a slightly different history with regard to how costs spent on newsgathering that could not be recouped were deemed necessary and important to the overall status of the network (for example, see Curtin on the value of Robert

Kinter's leadership for NBC news[18] or histories of CBS for William Paley's support of the news division). News divisions may not have been profitable throughout the network era due to the expenses of newsgathering, but profits otherwise flowed freely at the networks. When faced with profits inhibited by the mounting changes of the multi-channel transition, however, the tolerance for escalating losses by news divisions was quickly put to an end and new owners were less willing to accept the exceptionality of news.

Other adjustments developing during the multi-channel transition also challenged network news norms. The news divisions were still reeling from the evolution forced by new corporate owners when the 24-hour cable news channels began to mount formidable competition. CNN launched in 1980 and spun-off Headline News in 1982, but cable news didn't seem much of a competitor until the Gulf War in the early 1990s. Faced with the first major media event since the trimming of personnel and budgets, the broadcast networks then realized the consequences of their losses, while CNN took advantage of its full program day to provide constant and immediate coverage as well as its scrappy upstart culture that reinvented newsgathering techniques.

Cable news outlets, as a source for immediate video coverage of stories that did not warrant interruption of regular network schedules, certainly affected news culture—including adjusting viewers' expectations regarding what might be considered news and how news should be covered. Throughout the multichannel transition, the 24-hour cable news channels were able to expand the news agenda with endless airtime and offer frames that more precisely targeted particular audience groups—most clearly evident in the fast ascension of the FOX News channel, which joined an increasingly fragmented cable environment in 1996. NBC also attempted to compete in this space with its cable endeavors and partnerships, CNBC (1989) and MSNBC (1996) respectively, which received some support from the network's news division.[19] Cable news allowed audiences to choose among multiple channels that provided news around the clock. Such channels created the opportunity for television news to be timelier while also creating an unbearable burden of hours of airtime to fill. Yet, in examining the declines in network news viewing over the past 25 years, the audience losses don't appear to be linked to any explicit cable event. Combined viewership of broadcast evening newscasts declined steadily, with a few years of more pronounced drops and a few instances of slight gains. The years of audience losses greater than 5 percent (1987–8: 6.65; 1994–5: 8.22; 1997–8: 7.49, 2001–2: 8.23; and 2004–5: 6.25) do not coincide with changes in competition from cable or even notable news events. Also, cable news viewership more than doubled between 1999 and 2000, with no significant effect on broadcast nightly newscasts.[20] Although viewers were likely to tune in to the cable channels when desiring coverage of breaking news or when seeking updates throughout the day, the

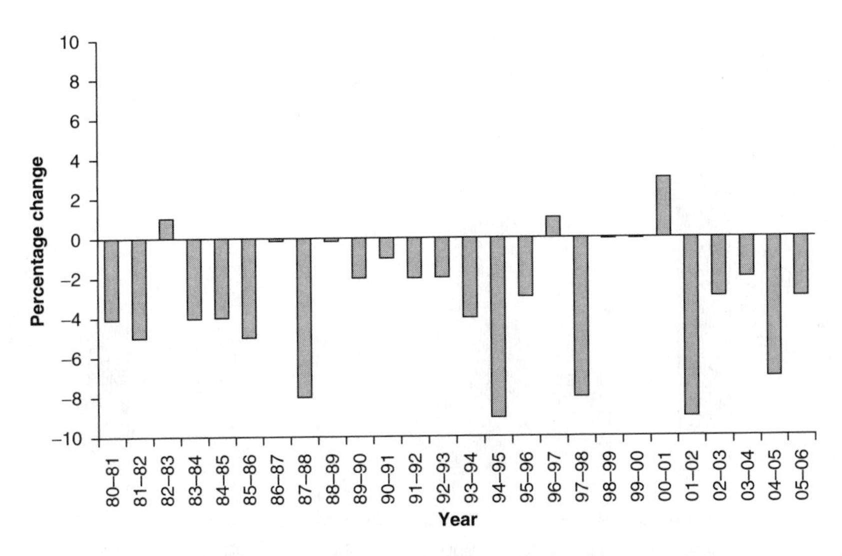

Figure 5.1 This chart illustrates the annual percentage change in aggregate viewing of nightly network newscasts.

Source: Compiled based on Nielsen data published at http://journalism.org/node/1344 (accessed October 15, 2008).

nightly newscasts remained part of ritual and routine catch-up at the end of the day's affairs. Cable never replaced this function for a sizable portion of the audience, nor did it directly pull viewers away at this hour.

Varied evidence indicates that a feature so mundane as ritual may go a long way toward explaining shifts in nightly news viewing. First, some in the industry recognize that any break in the pattern of nightly viewing can lead to audience losses that are never restored. Although his account is perhaps apocryphal, long-time network news producer Jeff Gralnick notes that disruptions, such as the weeks given over to coverage of the O. J. Simpson murder trial, led to some of the news viewing declines.[21] Indeed, one of the sizable year-to-year audience drops occurred between 1994 and 1995, which would include the period in which much regular network programming was adjusted to allow for coverage of the Simpson trial.

The second way the impact of ritual is being felt results from younger generations that are not developing the routine of nightly news viewing. Douglas Ahlers and John Hessen examined data of online and television news use and assert, "generational news-consumption patterns are of far greater significance to the well-being of the future of the industry than competition from the Internet."[22] Based on their review of data collected by a Magid study, they found that it is much less the case that viewers of television news are switching to online news

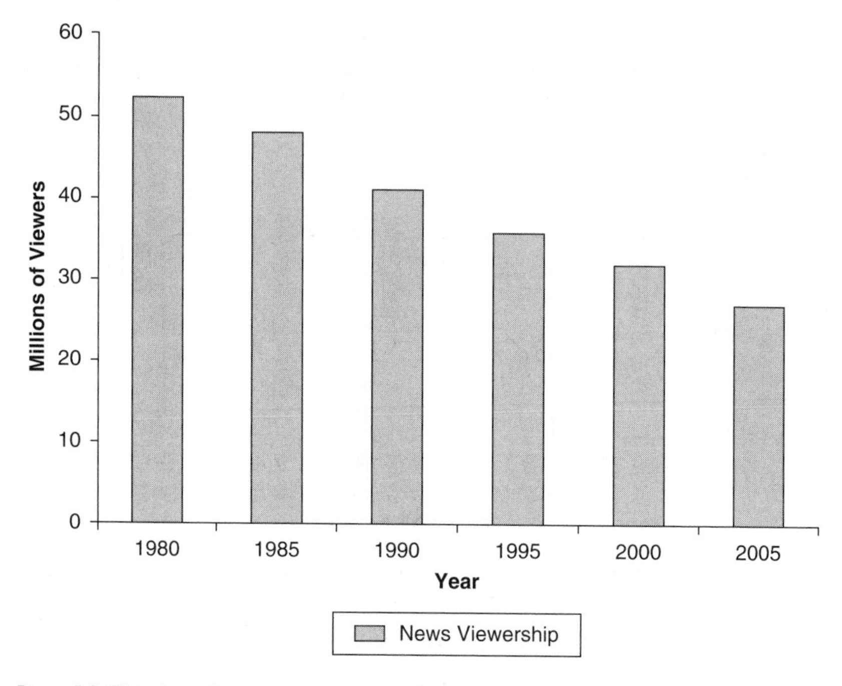

Figure 5.2 This chart illustrates the steady decline in viewing of national evening newscasts.

Source: Drawn from Nielsen data published at http://journalism.org/node/1344 (accessed October 15, 2008).

sources, but that the particularly old network news viewing audience—with a median age of 60.4 in 2006—is not being replaced by younger viewers who follow the same evening rituals.[23] As Jarrett Murphy noted in 2005,

> The 6:30 p.m. time slot might have worked in the '60s, when more families tended to gather around the set at that time. But the average American commuting time is now 26 minutes (in some big media markets it's longer), and Mrs. Cleaver isn't at home with dinner waiting any more—she's also racing home from the office.[24]

What Murphy acknowledges here is the wide-ranging social and cultural changes that reasonably can be expected to bear consequences for news viewing irrespective of new communication technologies.

Importantly, the arrival of cable challenged the network newscasts in other ways than by providing an additional outlet for news. The network newscasts likely derived some of their considerable share of viewership from the "roadblock" programming strategy of airing simultaneously. In the network era, viewers faced little program choice at 6:30, and watching television at that hour functionally required watching a newscast. Throughout the 1980s, the slowly growing number of homes subscribing to cable began to choose among other options during that time and the newscasts were forced to compete against non-news program forms. In the period from 1980 to 1990, 431 new local stations were also added, and these new stations provided alternatives to newscasts in the evening for those with only broadcast service.[25] The range of choices available only multiplied as the multi-channel transition continued.

Challenges and Opportunities for Network News in the Post-Network Era

The networks arguably never found a viable tactic to compete with cable news channels' 24-hour availability or with the non-news options offered by other cable channels and new broadcasters. They were, however, able to reach a sort of competitive coexistence after losing some initial audience share. Perhaps the key to sorting out the challenges that leave industry critics such as Friendly forecasting the death of network news from the true prospects that imperil the future of the form involves identifying the characteristics that particularly differentiate network news and its process of negotiating changes in industrial norms from other forms of programming. Cable competition was a problem of the multi-channel transition, but the post-network era and its consequences for television news present far more revolutionary disruptions and prospects.

One of the fundamental features of the national networks' nightly newscast has been its fixed time and related role in daily routines. Consequently, for news, the most crucial alteration of the post-network era may be the erosion—if not elimination—of regimented time norms in program scheduling. More than all other program forms, news derives value from the timeliness enhanced by the distribution possibilities that arrive in the post-network era. Its tendency to rely on short-form content further allows the control technologies of the post-network era to enable particular adjustments in a manner different from those that they deliver for other program forms.

Consider your own video news use and desires. These likely can be divided into two categories—a desire to know what is happening now, as in a rundown of the news of the day, and a desire to know more about something you already know is happening. The first function is adequately served by conventional newscasts only if they fit into your schedule. For example, if I want a sense of what important things happened today, I can expect that I'll get a decent sense of the highlights by watching a network's 6:30 newscast. But the limited distribution of linear content that was a key characteristic of network-era and multi-channel transition norms only allowed network newscasts to serve viewers if the audience could maintain the 6:30 appointment. This was a considerable inconvenience for the audience, but the lack of alternatives (that eventually develop from cable news services and Internet-based updates) meant that the network newscasts maintained their place. Conventional newscasts also rarely provide much to assuage my second category of news seeking—to know about something in greater depth—particularly if the topic is not of general interest.

But, in the network era, there was little that could be done to satiate viewers' desire for immediate news or greater depth of coverage. Nearly all news media—print, television, and radio—have featured "push" models of content dissemination that required audiences to wait until the next broadcast or publication for updated coverage. One of the more truly revolutionary advances of the Internet emerged from the manner in which its ability to republish inexpensively with great frequency created the opportunity for users to access the latest information. This was then augmented in the mid-2000s when bandwidth speeds made video distribution and sharing increasingly feasible (2006 was often touted as the "year of online video"). As such technological opportunities develop, television news divisions gradually have been incorporating applications that enable viewers to pull regularly updated information from their sites. This allows television news to be part of the adjusting norms of media use based on a "pull" model of content that are generally changing information dissemination. It also means that the networks no longer compete just with each other and cable news channels, but increasingly with all previously distinctive forms of news delivery, including "newspapers," "radio," and web-based news organizations, as all of these content creators are developing distribution applications that maximize the value of the web and Internet for dissemination.

The network-era characteristic of fixed-time newscasts limited both viewers and producers. Television news always had the possibility of breaking into regularly scheduled programming in the case of important news and had the advantage over newspapers of some sort of delivery multiple times a day, when factoring in the morning talk shows such as *Today* and the newscasts of local affiliates in the morning, at noon, and after prime time. Network news consequently was never that far behind a breaking story, but it lacked the immediacy that advantaged cable channels during a news event. A key adjustment of the post-network era then derives from the new ways of distributing video that eliminate any time gap between completing a story and circulating it to viewers.

A more technologically mundane post-network era device, the DVR, also offered easy time-shifting of news in a manner that dislodged it from its required 6:30 ritual engagement. People were unlikely to use DVRs to view nightly newscasts later in the week—in the manner that they might watch prime-time narrative programming—but DVRs could enable those who did not arrive home until 7:00 or later easily to maintain or re-establish rituals of nightly newscast viewing while preparing dinner or unwinding from a day at work. But while simply recording and later replaying content have proven so popular for narrative programming, the timeliness demanded by news makes the delivery platform of the Internet the more revolutionary development for this form.

The Internet—especially once it was able to disseminate video efficiently—required the networks to rethink their expectations of audience behavior, as online streaming and downloading jeopardized the linear model of scheduled news delivery while simultaneously providing the greatest tool for reinvigorating their businesses. Varied new distribution methods enabled viewers to become their own news directors and select the stories of greatest interest to create personalized newscasts. Online venues also provided tools for competing with the 24-hour news channels, as they allowed broadcast networks to keep information up-to-date and offer greater depth than possible when interrupting the regular schedule.

Because of the desire for timely news and the short form of its content, contexts that are considered suboptimal for other viewing—such as on computer screens and portable devices—are in fact ideal for news. Such devices and distribution enable place-shifting as much as time-shifting, so that the viewing of network news on the computer or on a portable device during times other than the 6:30 broadcast become far more significant adjustments than just delaying the linear broadcast.

Post-network era technologies and distribution possibilities, by which I mean primarily opportunities related to online video distribution that enable streaming and downloading to portable video devices, consequently expand the places and times viewers can engage with network news content. Although 6:30 viewership may be slipping from pre-eminence, post-network technologies create places and times to reach those lost and never found audience members as well. As a critical consideration, this is an important moment for reassessment of news practices not only because of the need to address how post-network operational dynamics affect the norms of television news. This reconfiguration of costs and practices offers a moment of change in which established practices are most vulnerable to adjustments that disrupt existing hegemonies about "how things are done" and what network news "should be," and the corresponding relationships of power that these norms enforce.

The enormity of changes in all aspects of television occurring as part of the transition to the post-network era consequently enables the reassessment of some of the old limitations for television news and creates a whole host of new possibilities. Particularly, the new distribution capabilities and technologies characteristic of the post-network era provide superior conditions for the production and reception of network news. Television news long has been valuable for its blend of image and sound and the timeliness possible through live broadcast. The convention of the network's schedule led news, although produced fresh each day, to be primarily cordoned off into a particular half hour that prevented it from providing the most timely update (unless the news was so significant as to warrant interrupting the "regularly scheduled broadcast") and required viewers to set their schedules around the newscast. Various post-network innovations assuage these limits by providing networks with tools to distribute constantly updated content and offering viewers ways to control when they view news and to select particular content.

Devices that expand distribution routes to allow viewers greater flexibility in when they receive news and the depth of content they can find are likely to offer the most profound adjustment of network-era operations. As of this writing, the networks' websites feature (or have featured) varying capabilities, including the ability to watch a stream of the telecast and/or a distinct webcast, à la carte access to the same video packages aired in newscasts and also longer versions, additional stories not part of the newscast, and the ability to download this content to other devices. These new ways for network news to reach viewers correlate well with the particular textual features of the form, such as its short length, and those characteristics valued in viewing news, such as timeliness. While old ways of distributing television news may be failing, the post-network era poses real opportunities for network news to advance beyond its previous confines.

The success of these developments are difficult to measure given the new and varied strategies with which networks continue to experiment and because measurement matrices are just emerging. The 2008 Project for Excellence in Journalism (PEJ) report on "The State of the News Media" included two notable pieces of data on this point.

1. MSNBC.com (the online portal for all of NBC news, including *The Today Show*, MSNBC, *NBC Nightly News with Brian Williams*, *Dateline*, and *Meet the Press*) ranked as the *second* most popular online news site in the United States, with an average of 29.2 million unique visitors per month. MSNBC.com trailed only Yahoo News, which gathered 32.6 million.
2. The ABC News online webcast draws a monthly audience of 4.5 million viewers (roughly half of how many watch the televised newscast on an average evening).[26] The webcast launched in January 2006 and is free, and can be downloaded from the ABCNews.com website or through Apple's iTunes. As of October 2007, ABC was the only broadcast network using its evening newscast staff to produce a separate and distinct webcast for an Internet-based audience.[27]

The MSNBC.com ranking is significant in that it reveals the complex and cross-media competition emerging. The other top nine online news sources include CNN (21.3), AOL News (21.0), Tribune (8.7), *New York Times* (8.6), ABC News (7.8), *USA Today* (7.4), and CBS News (6.9).[28] This reveals the transmedia competition involved in providing news in the online space. In contrast, the ABC webcast viewership data illustrate how small and insignificant the online audience is currently—particularly relative to the televised newscast.

In terms of production strategies, new digital distribution routes multiply the forms that the news product might take—in terms of both how it reaches viewers and what reaches viewers. NBC News president Steve Capus described the "operating idea these days" as "bring it in once and use it as many times as humanly possible, and that is how our newsrooms are set up and ... how the business is set up."[29] The newsroom and resource reallocation these comments suggest indicate that news divisions decreasingly see digital distribution and applications as ancillary. This fits well with the guidance that newsroom consultant Steve Safran advises. He notes, "Stations have to stop thinking of themselves as TV stations with Web sites. They have to see themselves as digital newsrooms. News goes in, news goes out, and when you gather something it immediately goes out."[30] Such a strategy diminishes legacy entities such as the fixed-time newscast, although it does not make it irrelevant. At ABC, the ABC News Now group delivers stories of various lengths to television, the web, and mobile devices and has deployed seven "digital reporters" around the world who focus on delivering content for multiple venues.[31] This transition requires wide-ranging changes in newsgathering and operations and allows for greater content variation.

News producer Jeff Gralnick, whose nearly 50-year career included stints at all three networks and at CNN, explains the networks' multiplatform endeavors as a last effort to maintain the economic feasibility of news production. Gralnick, who himself sounded the network-news death knell in a 2002 *Columbia Journalism Review* editorial, was less ominous in an interview upon his retirement in 2008, saying, "Television news as we know it is going to evolve into a totally new form or disappear."[32] Similarly, *ABC World News* executive producer Jon Banner confirms that the future for network newsrooms is not an either/or (television/web) proposition, noting, "We have to be doing both extremely well. The fact of the matter is at the moment our broadcast dwarfs any audience that we get on the Web, and probably will continue to do so for the foreseeable future."[33] ABC's strategy is geared toward a coming competitive environment in which the younger audiences of today maintain the rituals these new technologies allow for and tip the balance away from previous norms.

Although it is clear that network news divisions now understand that they need to embrace digital distribution rather than fearing cannibalization, it remains unclear what consequences this will have for the content of the news they produce. NBC News's Capus acknowledges, "The days of just doing the headline-news recap of the day are over, because people are bombarded with the news of the day. They know what's going on. We think people come to the *Nightly News* to find out why something happened."[34] In many cases, national newscasts may have pursued this strategy for some time. Andrew Tyndall, publisher of the weekly broadcast television newsletter the *Tyndall Report*, acknowledges that this "is a period of enormous innovation, but it's going to be innovation in delivery rather than content."[35] The televised newscasts face a notable challenge if their mission remains to create a one-size-fits-all newscast in an era of niche providers and narrowcasters. Altering the telecasts requires negotiating different approaches with a core audience accustomed to long-familiar norms. Consequently, experiments that substantially alter the format of content remain perceived as infeasible because they seem too likely to alienate the remaining older viewers.

But the telecast is no longer the only access to viewers, and ABC's still nascent *World News* webcast offers a good example of the content variation that another way to reach audiences can provide. Although the show is conceptualized to last fifteen minutes, senior producer Jason Samuels acknowledges that he doesn't "have to count the seconds," giving the newscast flexibility lacking on television.[36] The webcast does not have integrated commercial blocks, which also structure and limit the timing of television newscasts, and is free to include segments far longer than air on television. Samuels also pushes his correspondents and producers to deviate from broadcast news conventions: "Do one long standup, do much longer sound bites, play an interview. Produce a story in any way you think is engaging—there are no rules."[37] ABC executives acknowledge that, with the style and content of the show, which Brian Stelter describes as "an entirely different animal, sometimes resembling a younger, more technologically advanced version of the traditional 6:30 report,"[38] they are aiming for a 25- to 54-year-old audience who might view on iPods or mobile phones.

Many of the networks' efforts to update their distribution and monetization strategies for news through post-network technologies remain nascent and experimental. It is certainly the case that patterns of use have not shifted dramatically enough to consider emerging capabilities and uses dominant or mainstream, although patterns of technological diffusion suggest this someday might be the case. Yet it is already possible to consider the meaningful ways in which existing experiments with television news delivery alter the competitive environment in a manner that allows us to look ahead, not to predict, but to prepare for new ways of thinking about news that will be needed.

How might the on-demand, pull-based use of network websites change television news? This distribution method solves the problem of both lack of immediacy and lack of choice that limited network-era news delivery. Viewership expands because audiences can pull the very same stories from network news websites that others watch in the linear broadcast, but whenever and wherever they have web access.[39] In this viewing application viewers can choose only those stories of interest and effectively be their own news producer, watching packages and stories that were not included in the pushed newscast. Access to longer versions of the pieces that air in the newscast enable viewers to retrieve a greater depth of information, and, if they desire yet more information, it is easy quickly to follow links offered by the network as well as to search other sites. Even those seeking the familiar experience of pushed news can watch complete newscasts on network websites, download these stories to portable devices for viewing or listening during commutes home, or use DVR technologies to time-shift newscasts to a more convenient hour.

Despite the features of pulled video news that enhance networks' ability to meet people's varied needs, there is nothing in the technology or distribution possibilities to indicate that this will necessarily provide for any qualitative change in what is offered. Viewers encounter greater access to everything from stories of international importance to fluff celebrity "news." With greater control over story selection, an individual might compose a half-hour of news viewing filled with hard news stories—a more newsworthy option than the network selected newscast is likely to be on most nights—but also might construct a half-hour personalized newscast with no "hard" news at all. Networks aren't reporting what viewers are watching online, but it is reasonable to expect that they will adjust their newsgathering focus based on trends in audience preferences. Despite the technological shifts, the PEJ report revealed surprisingly little change over the last 30 years in what airs in the actual 6:30 newscast. The study, which compares newscast time spent on stories that are categorized as foreign affairs, crime, accidents and disasters, etc., frequently found that the content of newscasts as well as their format and style had changed minimally, leading Amy Mitchell, deputy director of the PEJ, to reflect that, "Network newscasts actually in many ways have been more consistent across the board than any other media in terms of staying with the traditional story approach."[40] What studies have not yet addressed is what stories viewers select to pull from available material and to what degree this differs from the pushed newscast.

In sum, post-network era distribution devices and technologies expand the ability of the networks to meet the varied needs of its consumers in a manner that is somewhat particular to news. Of course, much of this expanded capability comes through web technologies that are not the sole purview of the broadcast networks. Existing competitors such as cable news are just as capable of offering these new forms, and it is also possible for new competitors to enter this space—as has been the case. The well-reputed print news source the *New York Times* has added video stories to its webpage—although principally just feature stories—in a manner that suggests further breakdown in the boundaries among print and television news. Most local affiliates are also expanding their online news offerings, although these sites tend to focus primarily on local stories and on providing a local portal to news and resources. The Internet may make distribution of video news feasible for more entities, but for now still requires a significant infrastructure for video production that limits the main source of broadcast networks' competition to cable news networks and their affiliate stations.

Some may query whether distribution via the web qualifies these offerings as network television newscasts at all. Indeed, this is uncertain theoretical terrain at this point; however, regardless of the distribution mechanism, the boundaries of what we understand as "television" are still determined by an array of factors broader than the screen, device, or delivery system, but—in the words of Lynn Spigel—by a set of "technologies, industrial formations, government policies, and practices of looking" particularly associated with television.[41] It is not that "Internet news" can be conceived of as an entity separate from formerly dominant media such as print or television news, as, in many cases, the Internet functions as a distribution system, not as a medium in and of itself—at least within today's conceptualizations. Devices upon which users might view news are arguably currently involved in a struggle with our television *sets* over our "practices of looking," although the competitors of the television industry have aligned themselves with multiple viewing devices. It is possible that, as generations raised in the post-network era come to dominate society, we'll stop understanding video use as primarily the purview of "television," but this remains a justifiable and meaningful distinction for now. The technologies and practices of looking involved with post-network era news are highly differentiated from their network-era precursors. Nevertheless, they arguably remain those of television.

In addition to asking whether these newscasts are television at all, it is also reasonable to query whither the network in all this. Television networks historically have been defined as centralized entities that distribute programming to a system of affiliates. The circumvention of the physical components of this network-to-affiliate distribution system by post-network technologies does not negate the importance of the centralized entity we might still call the "network," even if it decreasingly functions in this way. The national economies of scale that led to the network system in the first place remain effective and even necessary to finance the costs of newsgathering, regardless of whether the affiliate system as it was originally conceived remains necessary for moving content from a central production site to viewers' homes throughout the nation. Certainly there is no business reason that the existing networks must be at the center of post-network television news, but their association with residual television experience provides a preliminary advantage.

Conclusion

Although it has seemed the case many times before, it may then be that the time has finally arrived to consider seriously claims about the death of the *network newscast*—which is not to assert the likely death of television or video-based news. The technological and industrial practices that are ushering television into a post-network era affect news in particular ways. More than anything, the new possibilities for video news indicate the importance of developing different measures for the vitality of network news than the audience share and advertising revenue of the 6:30 broadcast.

The limited time—roughly 20 minutes once advertising, promotional, and transition time is subtracted from the 30-minute telecasts—has always provided the form's greatest shortcoming in terms of both content and economics. As critics of television journalism have long complained, there is far more important news on any given day than can be fitted into the time available. The brief window of the linear broadcast requires network news producers to select among a vast range of content, a decision that is influenced by an imprecise calculus of what stories are most important, offer strong promotion potential, feature compelling video, or reach the audience sought by advertisers. Critics and scholars make cogent arguments about the negative consequences these economic and industrial features generate for the news that networks consequently make available,[42] as these are not criteria that often correlate with producing news likely to inform societies about meaningful issues relevant to the functioning of democracy.

Regardless of how networks managed these shortcomings in content, network-era news was somewhat inevitably destined to be an economic failure for the networks as well. It was, and remains, impossible to amortize the costs involved in maintaining a global and even national news corps in a single daily half-hour newscast. The particular economic circumstances of news production also distinguish it from other program forms. Newsgathering is costly, and news divisions have had limited opportunities to recoup or spread out these costs. In non-news genres, nearly all programs rely on repeat airings of some sort in order amortize costs and achieve profitability, or they are able to manage production costs in ways infeasible for news. Cost savings for other program forms can be achieved as a result of their predictability—attributes such as fixed sets and locations or the ability to record content in advance. The cost of maintaining a news force is simply greater than the revenue available from the approximately eight minutes of advertising available in each linear newscast.

The conventions of the network era—particularly pushing content at a set time with a significant delivery bottleneck—led to the narrowing of the news window, which challenged the economics of news divisions that would have been better off amortizing newsgathering costs across a longer program. A primary economic advantage of the cable news networks is that they gather a limited amount of coverage and then use it repeatedly throughout the day. Network news divisions only ever began to approach break-even figures and profitability once the vast news machine contributed more hours of content—particularly the prime-time news magazines. Plans to develop hour-long nightly newscasts emerged at various points in the form's history, but were ultimately not pursued because expanding the newscast would require taking back programming and advertising time from affiliates that was particularly lucrative for the local stations and unlikely to be returned—and certainly not without complicated negotiations and concessions.

Post-network era technologies and distribution routes present the opportunity to break open the constraint of the 30-minute news window. Yet, economics and distribution are tightly interconnected here, and just because new technological capabilities develop does not mean that the economics of commercial networks will enable their full use. The new distribution opportunities available to the networks provide tools to compete with breaking news coverage, but also contribute new and substantial costs as well. To compete in the post-network era, network newscasts have to find ways to leverage the content they develop with existing resources and to supplement those new distribution forms with new advertising revenues.

The preceding pages provide a narrative frame for understanding how and why the form of network news changed as it encountered the industrial developments of the multi-channel transition and the first throes of the post-network era. One of the greatest challenges of writing about various aspects of television's post-network era is the way in which it provides a constantly moving object of analysis. In the case of news, networks regularly have shifted the layout and range of offerings of their websites as their practices have remained highly emergent, experimental, and tethered to the constant evolution of norms of web use and technological possibility. Similarly, data about how viewers are responding to ways of accessing news remain limited and unrepresentative. As a result, any analysis of actual practice is likely to be unreasonably dated by the time it reaches publication, necessitating this broad-level consideration of post-network era news without actually knowing what "technologies, industrial formations, governmental policies and practices of looking" will come to dominate.

Identifying the economic model(s) that will support personalized news delivery remains the outcome most sought after by these industries and will have the greatest effect on the characteristics of the news and the access to it networks make available. The development of post-network era video news delivery remains limited by the sort of chicken–egg dilemma networks face in balancing the new technological capabilities they have with an uncertain economic model. To a large degree, networks are financing their experiments with post-network technologies and distribution forms with their revenue from their network-era newscast. Faced with decreasing—and aging—viewership of the fixed newscast, the networks are struggling to determine how they will finance their newsgathering once they can no longer rely upon the revenues of the 6:30 program. This underscores the manner in which the transition to the possibility of post-network-era television occurs within a system of constraints. The technology and distribution system may exist to bring about a much different set of norms for network news; however, the networks' need to ensure present and future profitability prevents the full incorporation of the possibilities these provide.

The centrality of the "network" in this assessment also belies hegemonic assumptions about the future of television that privilege existing entities—particularly the networks. While they possess a number of incumbent advantages, the scale of change within the television industry, journalism, and among all media is substantial enough that it is reasonable to question the persistence of the networks as dominant aggregators of video newsgathering and distribution. For now, their significance as perceived by the culture remains secure; however, this sense of distinction does not appear to extend to their corporate owners.

In 2006, Tyndall was quoted as saying, "The future of the half-hour nightly newscasts is inseparable from the future of the broadcast networks themselves. They will continue to decline at the same rate as the networks, no faster, no slower."[43] Tyndall isn't far off. Between 1980 and 2007 the combined nightly household viewership of newscasts declined from 28,459,900 to 18,954,400, a loss of 33 percent. Calculating the downturns in the broadcast network is a trickier prospect given the limited publicly available data and because season averages now include the ratings of FOX and the CW networks as well. A rough comparison suggests that the prime-time audience of just the Big Three networks has declined less than the 33 percent erosion experienced by news, as the entire broadcast audience (now including FOX, Univision, and the CW) has declined by 33 percent.[44] Another aspect worth consideration, though, is that the median audience age of news viewers is a decade older than those in prime time, which may further contribute to the sense of peril surrounding broadcast newscasts.[45] It may be the case that there will not be an evening newscast from three different broadcasters in the future, but it is narrow-sighted to dwell only on the size of the traditional 6:30 telecast audience and the revenue of this production as an indication of the health of network news. The network newscast can no longer be defined only by the 30-minute telecast, as it now encompasses a wide range of strategies that can re-establish the networks' status as newsgatherers and agenda setters.

Notes

1 Barbara Cohen, "Commentary," *The Future of News: Television–Newspapers–Wire Services–Newsmagazines*, ed. Philip S. Cook, Douglas Gomery, and Lawrence W. Lichty (Washington, DC: Woodrow Wilson Center Press, 1992), 34.

2 William J. Drummond, "Is Time Running Out for Network News?" *Columbia Journalism Review*, 25, 1 (1986), 52.

3 For simplicity, I simply refer to this as the 6:30 newscast throughout, although stations in the Central, Mountain, and Pacific time zones increasingly air this newscast at varied times.

4 Nielsen Media Research, *2000 Report on Television: The First 50 Years* (New York: Nielsen Media Research, 2000), 21.
5 Claire Atkinson and Bradley Johnson, "TV News Can't Carry Own Weight," *Advertising Age*, 17 January 2005, 3.
6 "Online Papers Modestly Boost Newspaper Readership," released 30 July 2006, http://people-press.org/reports/display.php3?ReportID=282.
7 Data drawn from Nielsen Media Research of 2006 audience figures. See www.journalism.org/node/1363, accessed 15 January 2008.
8 Nielsen data reported at http://www.journalism.org/node/1344, accessed 15 January 2008.
9 Atkinson and Johnson, "TV News."
10 See data reported at http://www.journalism.org/node/1339, accessed 15 January 2008.
11 Exact and even approximate financial data is extremely difficult to come by. Despite the networks being publicly held companies, profit and loss information is commonly reported across aggregate divisions, and even figures for news divisions include data broader than just the nightly newscasts.
12 Michael Curtin, "News in the United States, Network," *Encyclopedia of Television*, ed. Horace Newcomb (2nd ed., New York: Fitzroy Dearborn, 2005), 1653.
13 Michael Curtin, *Redeeming the Wasteland: Television Documentary and Cold War Politics* (New Brunswick, NJ: Rutgers University Press), 1995.
14 Joseph R. Dominick, "Impact of Budget Cuts on CBS News," *Journalism Quarterly*, 65, 2/3 (1988), 469–73.
15 Curtin, "News," 1655.
16 See accounts in Ken Auletta, *Three Blind Mice: How the TV Networks Lost Their Way* (New York: Vintage Books), 1992.
17 Although the affiliate stations received the licenses, the fact that the networks earned their most substantial revenues from the affiliate stations they owned and operated has always led to blurred lines between the affiliates and networks relative to the public service mandate.
18 Michael Curtin, "NBC News Documentary: 'Intelligent Interpretation' in a Cold War Context," *NBC: America's Network*, ed. Michele Hilmes (Berkeley: University of California Press), 175–91.
19 At launch, MSNBC was a 50–50 joint venture between Microsoft and NBC. NBC purchased a majority stake in MSNBC in December 2005, leaving Microsoft with 18 percent. Since its launch MSNBC.com has been a separate entity, and provides the web-based news portal for all NBC properties. MSNBC.com remains jointly held between NBC Universal and Microsoft.
20 "Cable News vs. Network News Viewership," http://www.journalism.org/node/1363; accessed 15 January 2008.
21 Jeff Gralnick, "How Network News Outsmarted Itself," *Columbia Journalism Review*, 41, 1 (2002), 54–6.
22 Douglas Ahlers and John Hessen, "Traditional Media in the Digital Age: Data about News Habits and Advertiser Spending Lead to a Reassessment of Media's Prospects and Possibilities," *Nieman Reports*, 59, 3 (2005), 65–9.
23 "Median Age of Evening News Viewers," http://www.journalism.ord/node1312, accessed 15 January 2008.
24 Jarrett Murphy, "The News at Dusk: The Death and Life of Network Evening News," *Village Voice*, 8 March 2005, http://www.villagevoice.com/news/0510,murphy1,61853,6.html, accessed 6 February 2006.
25 Jonathan Levy, Marcelino Ford-Livene, and Anne Levine, "Broadcast Television: Survivor in a Sea of Competition," Federal Communications Commission Office of Plans and Policy, September 2002, p. 19, http://hraunfoss.fcc.gov/edocs_public/attachmatch/DOC-226838A22.doc, accessed 9 November 2006.

26 Project for Excellence in Journalism, "The State of News Media 2008: Network TV," http://www.stateofthenewsmedia.org/2008/narrative_networktv_onlinetrends.php?cat=6&media=6; accessed 12 June 2008.

27 See Brian Stelter, "ABC Reshapes the Evening News for the Web," *New York Times*, 12 October 2007, http://www.nytimes.com/2007/10/12/business/media/12abc.html?_r=1&oref=slogin, accessed 26 October 2007.

28 http://journalism.org/node/1337. These figures are from 2006. I'm interested to see if there is significant adjustment (particularly increases in traditionally television outlets) now that video has become such an integral part of web use.

29 Brian Steinberg, "Those Stories Aren't Just for TV Anymore," *Advertising Age*, 28 April 2008, 12.

30 Daisy Whitney, "New Goals for a High-Tech News Market," *Television Week*, 14 April 2008, 26.

31 Ibid.

32 Michele Greppi, "Gralnik Observes 50 Years in News," *Television Week*, 28 January 2008, 4, 35.

33 Rachel Smolkin, "Hold that Obit: The Nightly Network Newscasts, Often Depicted as Passe, Face the Future with a Trio of New Anchors and Bold Plans for the Wireless World," *American Journalism Review*, 28, 4 (2006), 18–28.

34 Ibid.

35 Ibid.

36 Stelter, "ABC Reshapes the Evening News."

37 Ibid.

38 Ibid.

39 At the time of writing, this was true for ABC news, which was the most ambitious of the networks in their experiments with delivery. Most of the examples come from ABC news. NBC only enabled viewers to stream the complete 6:30 newscast three and a half hours after it aired, but also distributed it by podcast (both video and audio). CBS only enabled viewers to stream their newscast coterminous with its broadcast. See ibid.

40 See coverage in Toni Fitzgerald, "In Nightly News, Little That's Really New," *Media Life*, 20 March 2008, http://www.medialifemagazine.com/artman2/publish/Dayparts_update_51/In_nightly_news_little_that_s_really_new.asp, accessed 27 March 2008. The study does not address whether considerable change might be accounted for by coding adjustments.

41 Lynn Spigel, "Introduction," *Television After TV: Essays on a Medium in Transition*, ed. Lynn Spigel and Jan Olsson (Durham, NC: Duke University Press, 2004), 1–40.

42 Ben Bagdikian, *The New Media Monopoly* (Boston: Beacon Press, 2004); Edward S. Herman and Noam Chomsky, *Manufacturing Consent: The Political Economy of the Mass Media* (New York: Pantheon Books, 2002).

43 Meredith O'Brien, "Lost Cause? Network Executives Say Evening News Shows Remain Viable," *The Quill*, 94, (2006), 24–32.

44 News decline based on author tabulations drawn from data in the 2007 *Nielsen Media Research Report on Television* (news ratings figures for 1980 and 2007 multiplied by respective universe figures of 76,300,000 in 1980 and 110,200,000 for 2006). A rough comparison of prime time can be determined based on the 54.6 average rating of the Big Three for the 1980–1 season (Nielsen Media Research, *2000 Report on Television: The First 50 Years*, p. 18), which is equivalent to 41,659,800 households. A 33 percent decline from this figure is 32,078,046, which, coincidentally, is precisely the 29.1 average rating for all broadcast networks (including FOX, Univision, and the CW) in 2005–6, reported in the 2007 *Nielsen Media Research Report on Television*.

45 The median prime-time audience for all broadcast networks reached 50 in the 2006–7 season. "Primetime Regular Season Media Age Trends," *TVWeek.com*, 18 July 2007, http://www.tvweek.com/news/2007/07/primetime_regular_season_media.php, accessed 22 June 2008.

Chapter 6

Everything New is Old Again

Sport Television, Innovation, and Tradition for a Multi-Platform Era

Victoria E. Johnson

In the summer of 2008, the television trade industry magazine *Television Week* dedicated a special issue to "Sports: TV's Power Play." The feature essay stated that 2008 was "one of the strongest years ever for sports television."[1] Throughout the issue, sports programming's strengths are equated with network television's defining characteristics. Sport programming is "old" media, specifically "immune to the havoc wreaked by digital video recorders on other genres of programming."[2] Sport breaks through the perceived clutter of contemporary television by positioning traditional networks as home to its most iconic event telecasts. The "dependability" of networks as locus for sport programming underscores network television's continued relevance and establishes the key role of "sports entertainment as an instant brand maker for television outlets."[3] On the heels of a devastating Writers' Guild strike (2007–8), sport programming on network television apparently offers electronic, communal comfort food both to the business of television and to the broader culture. It epitomizes a symbolic return to a simpler, smaller, network-driven era of television characterized by "compelling TV" and "great storytelling" in a TiVo-proof, labor dispute-proof, and recession-proof, historically stable, ritually available package (characterized by sport's seasonal and calendar regularity).

FOX Sports president Ed Goren summed up this alliance of sport and network-era appeals best, noting:

> we invested billions of dollars in sports rights, ... with the belief that, as we move forward with more options for people, and as the television universe gets more and more diverse, the *one* segment of network television that would continue to be must-see TV and would continue to deliver large audiences would be the major sports events because that's where the water-cooler talk will be.[4]

Here, sport programming represents the last vestige of "mass" broadcast television with shared cultural value and import. Indeed, examples such as the Janet Jackson "wardrobe malfunction" at Super Bowl XXXVIII or rapt national attention paid to

Michael Phelps's 2008 quest for a record eight gold medals at the Beijing Olympics suggest that sport, as seen on TV, *is* the central, shared cultural forum for working through questions of community ideals, struggles over national mythologies, and questions of representative citizenship. Sport is "water-cooler talk," as a field of intellectual, social, and deeply *affective* resonance, recalling and epitomizing Horace Newcomb and Paul M. Hirsch's definition of television as an electronic public sphere or "cultural forum"—a key site for "the collective, cultural view of the social construction and negotiation of reality, or the creation of what Carey refers to as 'public thought.'"[5] Sport on television stands as a shared cultural realm through which we safely encounter and struggle with "our most prevalent concerns, our deepest dilemmas."[6] It is where "our most traditional views, . . . as well as those that are subversive and emancipatory, are upheld, examined, maintained, and transformed."[7]

While sport provides some of the most quintessential network-era programming, it increasingly *also* produces and engages in a context characterized by a "circulation of media content" that "depends heavily on consumers' active participation" within which "consumers are encouraged to seek out new information and make connections among dispersed media content."[8] In this sense, sport is, arguably, quintessential post-network-era content too. Unlike more traditional serial television genres, sport's "event" status and newsworthiness encourage its circulation in sound and audio bites and alerts across the multiple platforms of "new" media that are now part of every television network's corporate family. Whereas network-era address encouraged the "Monday morning quarterbacking" of water-cooler talk the day after a program aired, sport's desktop, laptop, and mobile content streams, in particular, now encourage *instantaneous* water-cooler talk away and apart from the television screen. The quick-hit, highlight-oriented, "instant" access appeal of post-network sport television—evidenced, below, particularly via online sites that are owned by traditional networks—is sport programming as, arguably, an increasingly *subcultural* forum. This is sport "television" that is characterized by individuated modes of address and à la carte packets of information and alerts delivered to the desktop, laptop, or hand-held mobile media device.

Indeed, as Goren points out above, FOX is "moving forward with more options" for content delivery within an increasingly diversified "television universe." In the same summer of 2008, the Beijing Olympics revealed that sport "television" would now be voraciously consumed and discussed *off* television, via platforms such as NBC.com's video streams and via mobile/wireless media. Coincident with the technological feasibility of online video distribution and a marked increase in US households with high-speed Internet access by mid-2008,[9] "NBC Universal created a new method of measuring audiences" combining "the number of television viewers with Internet users and people watching highlights on mobile phones."[10] The result was remarkable ratings, registering 206 million viewers in thirteen days.[11]

Thus, as the television industry adapts to new institutional logics and operational norms, sport programming epitomizes a paradox. It simultaneously maintains more network-era norms than any other program form, while it is also particularly well suited to new business practices, media outlets, and modes of viewer involvement that are enabled by the distribution flexibility and technologies characteristic of the post-network era. This chapter examines sport programming's historic and continuing synchronicity with network-era television practices in two primary ways: by tracing the form's centrality to institutional branding, differentiation, and claims to continued cultural relevance; and through the example of the foundational role of the National Football League (NFL) in shaping the business of television sport. Next, it examines sport television's inherent "hybridity" between network-era and post-network era audience appeals and modes of address.

Sport programming represents a symbolic and actual "bridge" between network-era practices and post-network realities. It represents a unique hybrid or *articulation* between network television's traditional role as the site of "mass" audience, communal, national spectacle (television as the home of larger-than-life events presented with the best view possible) and the post-network era's characteristic proliferation of content and its co-branded migration between media forms (from sport content's contribution to video-game aesthetics, to its foundation of online fantasy leagues and its key role in the development of mobile/hand-held video delivery technology). Sport's both/and qualities (as *both* epitome of network-era television *and* quintessential element of post-network developments) have assured sport programming's *unique* predisposition to anticipate, ameliorate, and capitalize upon the transition from the network era to the post-network era and beyond. Specifically examined here are sport programming's temporal and textual distinctions from other program forms. Sport's cutting edge visuality and aurality historically have been driving forces in television's aesthetic and technological developments. These qualities have simultaneously encouraged viewers'/consumers' continued engagement with sport as seen on TV and have also been important catalysts motoring new media adoption. To consider these issues, the chapter offers a case study of sport's foundational role in the post-network strategies and extensions of television networks through the example of CBS's CBSSportsline.com and the network's media rights agreement with the National Collegiate Athletic Association (NCAA). The chapter concludes with questions regarding potential dilemmas or limits to the paradoxical identity of sport programming and our engagement with sport across "old" and "new" media.

The Network Era, Institutional Identity, and Sport as "Brand"

According to communication historian Robert McChesney, the "sport–mass media relationship has been distinctly shaped by the emerging contours of US capitalism since the 1830s."[12] Prior to television, but coincident with the rise of broadcasting, "sport assumed its modern position as a cornerstone of US culture," as an "ideologically 'safe'" exploit that "lent itself to all sorts of civic boosterism."[13] Subsequently, with remarkable regularity across media history, times of economic crisis and technological transition have encouraged networks to stake their institutional identity on sport. Specifically, networks have "branded" themselves through their affiliation as the home of particular sports, sport leagues, and major sporting events.

In the radio era, network stalwarts NBC and CBS thus engaged in pitched battles over rights to broadcast boxing championships, which were "the most popular programs on the American airwaves."[14] And, when RCA publicly introduced television in 1939, two of the programs selected to highlight the wondrous capabilities of the new invention were a professional baseball game and a professional football game. Sport was also used to introduce television to curious but somewhat wary publics across the country during the 1940s and 1950s. Theater television broadcasts of Big 10 football, for instance, were especially popular in the Midwest, while heavyweight boxing matches brought viewers to such screenings around the country.[15] Theater television featured large-screen broadcasts held in movie theaters (the technology's key investor was Paramount pictures). Much like closed-circuit pay-per-view events held in arenas to this day, theater television typically featured sporting events, but also included newsworthy events broadly conceived (such as political conventions or speeches).

Television's earliest weekly series programming also incorporated sport. In the late 1940s, Wednesday night's *Kraft Television Theater* on NBC was "counter-programmed" by ABC with wrestling and by CBS with boxing. In the 1949–50 season, television owners could watch a sport series in prime time five nights a week.[16] However, sport on television—and network fortunes as they were pegged to sport—began to flourish in the 1960s, in direct correlation with the consolidation and burgeoning national success of the National Football League.

The 1961 Sports Broadcasting Act was critical to the explosion of television sports in the 1960s. This act, which made sports leagues exempt from antitrust laws, permitted professional sports leagues or conferences to broker broadcasting agreements for all of their teams as a "package" to a network or networks. Before this period, individual teams would each negotiate their own, local telecast deals within their own market. This put teams in small markets at a relative disadvantage and mitigated against any sport being perceived as truly "national" or any fan as being more than a fan of her/his local team. According to scholar Michael Oriard, when the NFL's new commissioner, Pete Rozelle,

> was elected in 1960, the league's 14 clubs had individual television deals ranging from $75,000 for Green Bay to $175,000 for the New York Giants. In 1961 Rozelle persuaded the most powerful major-market owners ... that short-term sacrifice would pay long-term dividends. Sharing television revenue meant rough parity and financial stability ... More important, ... because the NFL could never have franchises everywhere, viewers willing to turn on pro football every week in much of the country would have to be fans of the league, not just of the New York Giants or Los Angeles Rams.[17]

The NFL, as it is now known, did not exist at this point. Until 1966, the NFL consisted of teams that are now, in the main, part of the league's NFC conference (though the League has expanded exponentially since then). A competing professional football league, the AFL, was also viable in this period. In fact, in 1960 the AFL, perceived as the "secondary" pro league, and ABC, perceived as an "also-ran" broadcast network, forged an alliance—both, arguably, staking their institutional identity and national "brand" on one another—by signing a five-year, $8.5 million television contract, while CBS carried NFL games.[18] As the television audience for football grew (and, arguably, encouraged the development and proliferation of other sport programming), rights fees also began to increase. In 1966, the NFL and AFL announced a merger agreement (effective in the 1970 season) that created the "Super Bowl" and allowed the league to forge unified rights deals. The Super Bowl was, thus, initially, the annual championship match-up between the leagues and then, post-1970, between competing conferences within the NFL; its telecast has since rotated between each of the league's network broadcasters. The NFL settled on a split television package agreement, which, to a great extent, still exists today: CBS would air NFC games, while NBC received the rights to AFC games. In 1970, ABC successfully returned to football telecasting with what would become the blockbuster of all football coverage—*Monday Night Football*.[19]

Using proprietary league rights agreements to establish institutional credibility and solvency is a business practice that historically has been adopted by the "weaker," "disadvantaged," and "start-up" player who has, subsequently, been central to each historical transition in television networking. The risk ABC took in signing on with the AFL paid off in institutional stability and exponential growth for both entities. In the "gap" between its coverage of AFL games and NFL/*Monday Night Football*, ABC became the "worldwide leader in sports" by developing its expertise in Olympics coverage and weekly digest programming, such as *Wide World of Sports* and *American Sportsman*, as well as college football and college football bowl games. In another case, the Home Box Office network (HBO) began in 1972 as a subscription television service that distributed theatrical films and sporting events into paying viewers' homes. HBO "debuted on November 8, 1972, telecasting *Sometimes a Great Notion* (1971), starring Paul Newman, and a

National Hockey League game to a mere 365 cable-subscriber households in Wilkes-Barre, Pennsylvania."[20] And yet, in 1975, when HBO debuted its satellite cable service, "with the much-ballyhooed 'Thrilla in Manila' heavyweight boxing match between Muhammad Ali and Joe Frazier … In one fell swoop, HBO became a national network, ushering in television's cable era,"[21] and representing the future of sports after the network era.

Following HBO's contribution to sports on television, the multi-channel transition delivered further adjustment in close succession. The "Superstation" phenomenon—using satellite redistribution, effectively, to make a local independent broadcast station a national cable and satellite service outlet—launched and became viable largely on the strength of local sports programming. Ted Turner's WTBS began national telecasts in late 1976, originating from Atlanta and featuring Atlanta Braves games. Chicago's WGN launched nationally in October of 1978 and offered Cubs and Bulls games (and, later, White Sox games as well). Interesting in the context of post-network diversity noted above, WGN's prior institutional claim (to be the "home" of Chicago sports, uniquely available nationwide) has been revised. In a post-network era, WGN has reinvigorated network-era branding logic by promising breadth and "mass" appeal staked primarily on sport content. Specifically, in the summer of 2008, WGN rebranded itself as "WGN *America*," supported by the fact that it "carries more baseball games than any other national television network."[22] Much of the network's remaining schedule currently features "retro" series program broad-appeal favorites, including *The Bob Newhart Show*, *WKRP in Cincinnati*, and *The Honeymooners*. Nationally available sport programming from both Major League Baseball's American League (White Sox) and National League (Cubs), combined with the broad demographic-appeal programming of "Outta Sight Retro Nights," ironically position the superstation as a "throwback" home for network-era appeals.

Although sports coverage was crucial to the success of superstations on cable, ESPN has been the iconic channel of the multi-channel transition and sport to the present. ESPN, which launched in 1979, quickly became the most popular 24-hour cable outlet for sport programming. In the early 1980s, Capital Cities/ABC was in negotiation for its purchase, and by 1984 it was a majority owner of ESPN with the Hearst Corporation. So ESPN has, historically, been in the same "family" as one of the key Big Three broadcast networks, ABC.[23] In other words, ESPN's early and continued intertwining with ABC calls attention to the ways in which sport content and the business of television sport, historically, have connected network-era practices and industry "players" to post-network practices and entities. ESPN, arguably, indicates an early "extension" of the ABC network into a post-network era, and at present ESPN's fortunes and multi-platform diversification help subsidize ABC's operations. Today, ESPN represents the "ideal" of the multi-channel era transition in terms of its radically "niched" and diversified outlets. Because of its multi-mediated diversification and global reach,[24] it is also poised to continue to dominate as a media player in the post-network, multi-platform era.

Perhaps the historic icon signaling the centrality of sport television rights to both network television business practices and the institutional success of entities developing after the network era came with the emergence of FOX television in the late 1980s. FOX network launched in 1987 and staked its identity and institutional toe-hold on three key strategies: building up its audience by targeting typically underserved television viewers (youth and African-American audiences in particular); scheduling new programming during the otherwise rerun-ridden summer months; and successfully outbidding CBS for a portion of pro-football broadcasts each season.

In 1993, FOX offered an unprecedented $1.58 billion to the NFL for rights for which CBS had offered only $290 million. FOX's overwhelming offer was possible because it *was* a non-traditional network. It did not have the "Big Three" networks' sports or news divisions with their expensive personnel or infrastructure. It did not offer a full, seven nights per week prime time programming schedule and, thus, its series development costs were considerably lower.[25] And FOX had the relatively unlimited resources of Rupert Murdoch's News Corp., perhaps the signal media conglomerate ushering in the post-network era.[26] Its rights to the NFL allowed FOX to become a legitimate fourth network by adding millions of weekly viewers, affiliates, and marketing targets for its new prime-time series throughout the week. Thus, as an emergent network, FOX is characteristic of the multi-channel transition, but it only flourished as a serious threat to traditional networks by literally and figuratively capitalizing upon NFL football's familiarizing, "mass" network-era appeals. A major sport broadcast rights contract, iconic of the modern network era, was thus necessary for the post-network transition to truly begin.

From the 1990s to the present FOX has aggressively challenged the sports centrality of Disney Co.'s ABC and ESPN, particularly via its regional sports networks (RSNs). RSNs are cable outlets that carry the rights to telecast local sports teams' games. These outlets obtain the local cable rights to telecasts that typically also have local over-air broadcast carriers. FOX Sports Net is one of the largest RSNs and recently won a rights battle with ESPN that blocked the latter's attempt to start an RSN in Southern California.[27] Cable service provider Comcast is the second largest RSN operator. Smaller RSNs (which yet serve very large markets) include, for example, New England Sports Network (which is co-owned by the Boston Red Sox and the Boston Bruins) and Yankee Entertainment Sports.

Recently, the NFL got into a legal dispute with several of cable's multi-systems operators, or MSOs, because each refused to add the NFL network (2003–) to their basic cable line-ups. MSOs such as Comcast, Cablevision and Time Warner each own multiple cable television franchises and deliver communication and entertainment services beyond cable channel program offerings (including on-demand video services, high-speed internet, and telephone services). Each of these cable providers argued that the league's new network belonged on a "sports tier" where customers would pay a fee for a cluster of special-interest channels without passing additional subscriber costs on to basic cable subscribers. The NFL argued that these cable operators were protecting their own regional sports outlets (even though none of these outlets are licensed to broadcast NFL games) rather than defending basic cable subscribers.[28] In the Comcast case, for instance, the NFL's argument was that, by packaging the NFL network with a "sports tier," cable operators with regional sports networks would be using the NFL brand to entice viewers to purchase additional sport programming that would directly profit the cable operator while simultaneously costing the NFL "basic" cable viewer/subscriber fees. The NFL network would "sell" the RSN-populated tier to viewer/subscribers, in other words.

Rhetorically, the NFL fought plans for "tiering" by arguing that its new cable outlet actually epitomized the ideals of traditional network-era broadcasters. It resisted the idea that its network was "niche" viewing, proposing instead that sport programming is "public interest" programming that should be carried on "basic" cable line-ups because of the broad appeal, cross-country geographic interest, and shared civic engagement of NFL games. The NFL upped the stakes of this argument and promoted its "football telecasts are in the public interest" argument by joining its "over-air" partners in broadcasting key end-of-season Thursday night match-ups. Those viewers without the NFL network were, thus, now missing games that they "normally" would have viewed via traditional over-air broadcast.

All parties here engaged in rhetorical play as regards notions of media access, democratic "rights," and public interest programming. The NFL, for instance, has a special subscription package with *only* DirecTV for its "NFL Sunday Ticket," through which viewers can watch every NFL game on any given weekend. Needless to say, direct broadcast satellite service—or in-home, satellite-delivered television services such as Dish Network and DirecTV—is not broadly "shared" programming because, like cable TV, it requires viewers to have the resources to subscribe to its packages of program service. Adding NFL Sunday Ticket to a basic DirecTV subscription, for example, can cost over $250 for the season. Further, DBS services are direct competitors of MSOs such as Comcast, who thus claim that the NFL network is a service that is more analogous to the NFL Sunday Ticket package than it is to the NFL's weekly broadcast homes, FOX, CBS, or NBC. Considering its historic clout and centrality to television's solvency, however, the NFL's voice in such disputes is one that overwhelmingly compels compliance.

In *Brand NFL*, Michael Oriard writes that "The marriage of the National Football League to the television networks has been the most intimate and mutually enriching in American sports."[29] For telecast rights agreements that run through the 2011 season, CBS has paid the NFL $622.5 million; NBC has paid $600 million; FOX has paid $712.5 million; and ABC's sibling, basic cable sports giant ESPN, has paid $1.1 billion to the league (for rights through 2014).[30] While much has been written regarding the outrageously exorbitant nature of these fees, "afternoon NFL games" consistently "clobbered the prime-time average ratings at all of the networks (by 20 percent for CBS and 65 percent for Fox)."[31]

Indeed, losses and gains in sport television programming and its profits tend to be highly cyclical in nature. Historically, big gains are shown when there is increased demand for new, unscripted, original programming, and, in conjunction with increased adoption of new television technology (most recently exemplified by spikes in ownership of HDTV),[32] particularly when coincident with sporting events of major national, historic interest. The 2007–8 sporting year is an example of a perfect storm of sport television success in both of these terms as demand for original content dovetailed with increased availability and demand for new media technologies predisposed to "show off" sport. The Writers' Guild of America strike in fall and winter of 2007–8 led to a pent-up demand for "fresh" television. In the NFL, the 2007–8 season featured the compelling narrative of the New England Patriots' quest for a perfect record, which culminated in a loss at Super Bowl XLII, but drew the "largest average audience in the game's history at 97.4 million."[33] The summer of 2008 saw increased access to broadband technologies in American homes (enabling more viewers to access broadband video online) and witnessed continued growth in HDTV ownership, both at the same time that NBC showed the 2008 Beijing Summer Olympics across its television and online universe. Journalists Scott Collins and Lynn Smith noted that viewing of this Olympic Games exceeded "the audience for most airings of Fox's musical smash 'American Idol,'" a feat they note as "especially impressive considering that, thanks to out-of-town vacations and daylight stretching past 8 p.m. in some areas, fewer people watch television during the summer months than at other times of the year."[34] Thus, television networks that are in the business of bringing sport to the US viewing public actively promoted themselves as inherently "democratic" institutions and protectors of revered cultural and seasonal traditions, as they featured programming that joins "mass," broadly mixed demographic communities and delivers ritual, historic events such as the Super Bowl and the Olympics. Sport television was the familiarizing framework through which new delivery technologies (e.g., HDTVs) and new platforms for engaging television "off"-television (e.g., via broadband and wireless Internet connectivity and mobile media devices) were adopted. Sport program content, audience address and appeals thus represents the articulating term between network-era traditions and post-network innovations.

The "Pre"- and "Post"-Hybridity of Sport Programming

In spite of the critical core network-era attributes of sport television, sport programming has *always* challenged conventional ways of thinking about the medium through its scheduling, modes of address, and audience appeals. Within the business of television, sport is "hybrid" in terms of scheduling, temporality, its formal properties and genre identity, and encouragement of viewing rituals that are often communal and public rather than domestic and intimate. In terms of temporality, sport programming is uniquely "unpredictable." Its key texts exist, most typically, outside prime time or overlap and stretch prime time's borders in unpredictable ways. Formally or textually, through both visual and aural means, sport programming has regularly been *the* site through which *new* modes of television technology, aesthetics, and address have been introduced, and by which viewers have become familiar with and been encouraged to adopt new technologies and applications: from color receivers to HDTV, from portable "Watchmen" to wireless enabled cell phones, sport is both the epitome of network-era spectacle and communal "event," as well as ideally predisposed to being parsed out in small "bytes" of information or highlights and news alerts best suited for miniaturized technologies and à la carte delivery. Every sport program is additionally hybrid in that each crosses multiple generic boundaries (each exhibits characteristics of the "spectacular"/dramatic narrative or compelling contest of game shows, while also indebted to news, public service programming, and talk shows). And sport programming's coverage and appeal also often remain intensely local and regional in spite of television's presumptively national address in an increasingly global media era.

This chapter proposes that, together, these "hybrid" qualities of sport programming, historically, have *both* made it the epitome of network-era television at its zenith *and* have encouraged networks faced with economic and technological upheaval to capitalize on sport programming's inherently *multi*-mediated, individually directed properties that are central to post-network applications and success. It is this doubling of the communal, "mass" audience, shared cultural experience, *and* personalized, individuated, "à la carte" potential of sport that defines its centrality to the continued market strength and cultural significance of network television *and* to the promotion and realization of the post-network era's multi-platform model of "networking."

To unpack this paradox, first we consider sport programming's *temporal* distinctiveness, within and apart from serial television flow. Sport television's temporal play is reflected in two primary ways: as a characteristic of its scheduling and as an internal, audio-visual property of sport programs. Beyond the borders of its program "texts," sport has a distinctively unpredictable status within television's schedule. Televised coverage of sport means, most often, covering an event live, in "real time," which can wreak havoc on "regular" televisual "flow."[35] As noted above, sport's temporal unpredictability has allied it with traditional, network-era sponsorship and audience-gathering practices: sport programming "frustrates" new media applications such as the digital video recorder because it continues to command large, diversified audiences in *real* time (DVRs capture programming that begins and ends on the hour, whereas sport programming often overflows its scheduled time period, thus rendering "regular" TV time incompatible with sport time). Sport programs thus represent an exception to the post-network television temporality that tends to be associated with viewer-initiated "time-shifting" practices (originally ushered in by the VCR, and now epitomized by the DVR and phenomena such as "on-demand" video). In this sense, sport's temporality recalls network-era viewing practices and the zenith of the Big Three networks as destinations for "appointment" viewing and shared viewing experiences, across the country, in simultaneous time (e.g., in spite of time zone differences, the West Coast and East Coast watch the Super Bowl at once). Sport's temporality and flow, as seen on television, particularly when at its most "event"-worthy, thus restore the medium to its network-era identity as a venue for simultaneous connection across demographic communities and time zones and as a site of shared culture experienced in real time.

Within sport programming, slow motion and replay further confound "normal" concepts of temporality. Televised sport is characterized by "spatial compression and temporal elongation and repetition" via use of "extremely long lenses" which effectively flatten space and distort the field of vision.[36] Sport allows its viewing audience to break down time and space, making the previously invisible visible and granting "enhanced" sight to each viewer. Technologies such as the telestrator (which allows writing on-screen by which plays and strategies can be mapped out for the viewer), the "virtual first down" yellow line (which marks remaining yardage required from the line of scrimmage), the Skycam or Cablecam (the camera suspended over the field of play that allows for "north–south" orientation to the field rather than just the "east–west" orientation that mimics the view from the stands), and the T-Mo, an extreme slow-motion hi-def camera (which is capable of capturing over 1,000 frames per second) guarantee the television viewer a better perspective than those attending in person. Watching sport live on television is, ironically, better than being there. In this way, visual technologies and audio advancements, including in-game players who are wired for sound, capitalize upon and epitomize traits that historically have been hailed as *specific* to television. Particularly, televised sport is a site of "mobile privatization" *par excellence*. The home viewer in "privatized"—or private—domestic space is transported to the site of sporting events, or made mobile. Once "there," she or he is given a significantly enhanced view on every angle of the action. Here, sport television capitalizes upon and reinforces what *television* can do that other media *cannot*. Sport exemplifies television's very "televisuality" for a new media era. Unlike most any other television genre, sport cannot be imagined *without* television, and vice versa.

Sport's unique televisuality does not preclude its migration to other media (as will be demonstrated below), but it does reinforce the genre's centrality to the medium and its distinctiveness *within* television program address. According to

Margaret Morse's foundational "Sport on Television: Replay and Display," sport programming represents a unique generic blend of news and public service programming with spectacle and epic drama. Sport is "news"-like in its structural hermeneutics (each game or match is a contest with the question to be resolved: who will win? The outcome or answer is, then, unquestionably "real" in the same way that we apprehend news to inform us of historical facts). Sport television is, additionally, overflowing with statistics (expressed both aurally, via "color" commentary, and visually, through on-screen graphics) that are presented in "real" time while contextualized through a historical and a future-looking perspective. Our engagement with sport thus always seems uniquely newsworthy *and* larger than life. Writes Morse, the "aura of scientificity of sport, its news-value, and its perceived realism protect its *extraordinary* status."[37]

When developing the original *Monday Night Football* (ABC, 1970–2005; ESPN, 2006–), Roone Arledge proposed that televised sport should *be* extraordinary in all of the above ways. Arledge argued that sport programs should place the viewer in the center of the action, representing a distinct departure from the experience of attending a game in person, in the stands. The better view television could provide would, Arledge proposed, fuse "art and journalism," or narrative drama and newsworthy documentary "in a way that was totally compelling."[38]

Historically, sport's aural and visual complexity and temporal play have been definitive of the "cutting edge" in television style. While other television formats—especially those occupying the schedule outside of prime time—are often, visually, fairly static, indebted to their genesis in radio, sport television represents television at its most aesthetically masterful. It is profoundly visual and technologically advanced, characterized by "the highest production values in regularly scheduled television," event television, and across the media industries.[39] New photographic, graphic, and temporal techniques unique to television (e.g., unavailable in cinematic or in mobile/hand-held technologies, particularly) are, generally, first seen in television sports coverage. Sport has regularly been *the* site through which new modes of television technology and address have been introduced and by which viewers have been encouraged to adopt new technologies and applications through comfortable, familiarizing screens (from color television to high definition). Thus, while a proclivity for watching sport "live" may have served as a roadblock to DVR use, sport fans have, conversely, driven the adoption of High Definition TV, positioning it as "the fastest-growing consumer technology segment in 2007."[40] Notes Natalie Finn of *Television Week*, "Enthusiasts looking to enhance their sports-watching experience are one of the largest forces driving HD television sales, . . . one in three HDTV buyers expects to watch more sports after his or her purchase, and 71 percent of sports fans up their intake after getting a hi-def set."[41] But how do these characteristics of sport position it as *both* the ideal network form *and* the ideal post-network extension?

Sport in the Post-Network Era

Broadly, in a context in which cable and new media platforms are increasingly "the driver of broadcast fortunes," sport's inherently cross-generic, temporally unique, hyper-textual, multi-mediated public and private identity has encouraged and insured its successful "microcasting" across the platforms of each of the "traditional" broadcaster's media "families."[42] Overall, the history and projected future of sport television remains one that I term as "tradition within change" and of the ideological power of broadcast networks as "shared," "democratic" media space, even when they are engaged "off air," online, or "in hand" via venues that, in fact, encourage and complement viewers' return to the television screen.[43] In the post-network era, traditional broadcasters (ABC, NBC, CBS, and FOX) use proprietary league/conference rights agreements and sport content particularly effectively to create and project coherent brand identities—identities that create seamless content and movement between television broadcasts and the networks' online and mobile media investments. Now that "more Internet users are capable of watching video streams than ever before," evidence suggests that the Internet is having a powerful complementary effect—increasing "both [viewers' engagement] with TV and their use of it."[44] Indeed, shown here in the example of CBSSports.com, post-network multi-platform venues now enhance and supplement *traditional* telecasts with real-time updates, video streams, and mobile alerts that capitalize fully on televised sport's "newsworthiness" and "realist" appeals while allowing for post-television media's individualization of the fans' engagement via personalized information streams.

Such post-network-era sport content and its modes of engagement are exemplary of "flexible microcasting." Scholar Lisa Parks uses this term to describe the "industry's visions of a new kind of *personalized* TV" seen "in 'postbroadcasting' strategies that combine historical, over-air, cable and satellite programming and flow *with* computer technologies to reconfigure the meanings and practices of television."[45] These practices are characteristic of the multi-mediated economic and creative sites of US network investment (or multiple "platforms") both on and off television (for the PC or mobile device, for example); these investment strategies and applications currently complement and, often where sport is concerned, actually *expand* traditional network viewing and engagement levels, as noted above, and exemplified through CBSSports.com below.

Beyond sport *on* television, then, networks have invested deeply in sport as a leading multi-platform content application—financially, technologically, and conceptually. Such applications extend network sport television's presence to mobile phone alerts, WPA-enabled streams, and online sites; they extend sport programming's appeals beyond network-era "communal" engagement to encourage individual interactivity and multi-tasking *with* sport; and they directly appeal to the individual viewer with strategies that emphasize "à la carte" information addressed

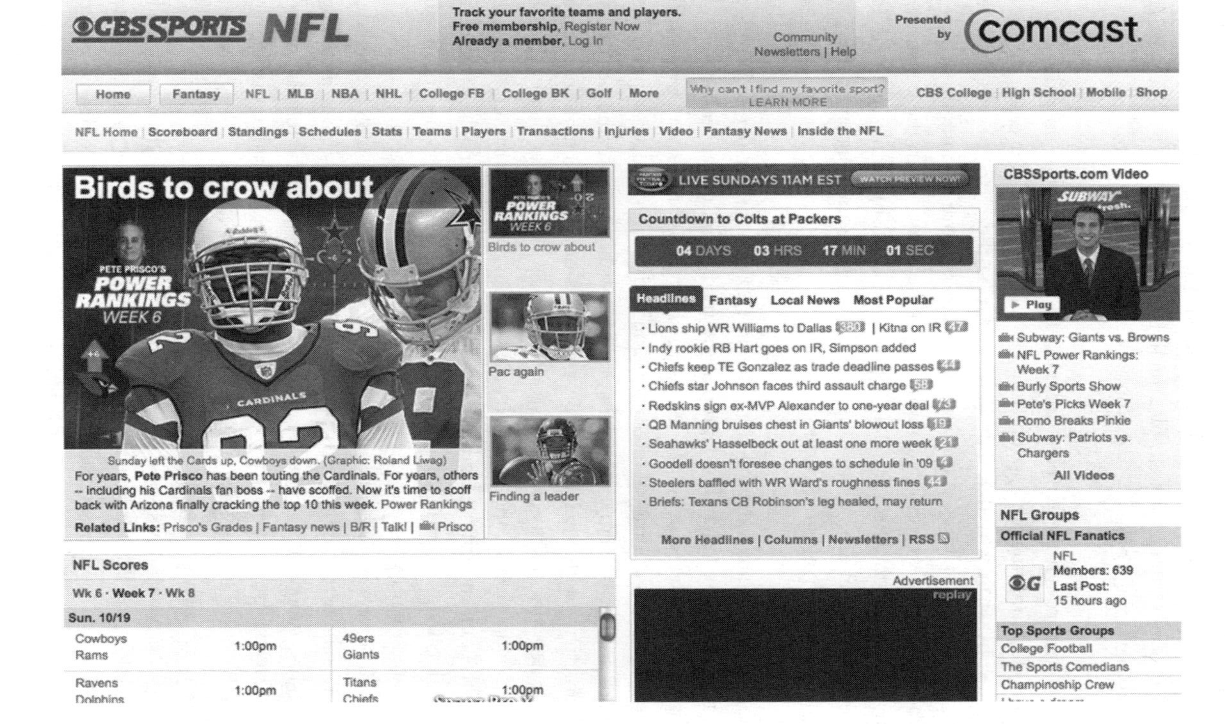

Figure 6.1 This image of part of the CBS Sports NFL website illustrates the range of information, video, opportunities to interact, and advertising aggregated on the site.

Source: http://www.sportsline.com/nfl (accessed October 15, 2008).

to individual fan passions on a "micro"-scale, in everyday use.[46] These applications of sport "television" capitalize on sport's dual ability to be a "larger than life" communal spectacle, as traditionally seen on TV, *and* to be individuated and integrated into new technologies that are, now, always at hand. Superstation WGN (television home of the Chicago Cubs), for example, might be partners with a mobile service provider and marketing company that would work with sponsors to

> collect data through the wireless Internet to determine not only the exact location of a consumer at a given time (at Wrigley Field, for example) but also the context of why that individual might be there (to cheer a favorite team, the Chicago Cubs). With that information, more meaningful or relevant advertising messages or promotions can be delivered to the consumer (a 30% discount coupon for select Cubs merchandise) on his [*sic*] mobile phone or other hand-held device in a setting where the consumer may be receptive to receiving such communications.[47]

In this example, whether watching the Cubs in HD at home or live at Wrigley Field, the mobile phone in hand or laptop at the ready can further "enhance" and "personalize" sport television off the edges of the television screen, while still serving the interests of traditional telecasters and their multi-media partners. The post-network multi-platform era and its media collaborations combine a new level of interactivity with a heightened degree of location specificity to converge with historic network television interests, rather than overthrow them. Mobile technologies, online access, and fantasy leagues encourage the sport fan to engage with her or his team and fan community virtually, no matter where she or he may be physically located, suggesting that sport remains a field of everyday, localized, and individuated identity as much as it is spectacular and communal.

For a specific example, CBS offers a useful case study. In 1999, CBS brokered a deal worth $6 billion to claim ownership of a package of NCAA rights, including broadcast rights, cable outlet rights, satellite partnerships, digital television and home video rights, "plus Internet, electronic commerce rights, radio, marketing, sponsorship, licensing and merchandising rights."[48] While CBS's NCAA agreement included broadcast rights to "lesser-known," "Olympic" sports such as lacrosse and soccer, the centerpiece of the deal was the NCAA Men's Collegiate Basketball Tournament known as "March Madness." Starting in 2002, CBS began to integrate its NCAA broadcasts (both basketball and beyond) across all of its "company-owned outlets, as well as through partners such as DirecTV, the No. 1 satellite television service, and YouTube, the biggest video-sharing site."[49] In 2007, it was estimated that NCAA Tournament broadcast revenue *alone* would garner CBS over $500 million.[50]

Beyond the television screen, however, CBS's methodical cultivation of an online presence for NCAA sports, and particularly March Madness, has coincided with and benefited from growth in broadband high-speed Internet access and use. In March of 2008, business reporter Chris O'Brien noted that, "According to a recent report by Robert Peck, a Bear Stearns analyst, 75 percent of folks online now play videos, up from 60 percent a year ago. Those viewers spend an average of seven minutes a day watching video online," and, in March, "one of the biggest sources of that traffic was expected to be CBSSports.com."[51] With CBS's streaming video of all games taking place in or out of prime time, free, online, many employees are watching at work. Indeed, CBS developed the "Boss Button" for such settings that, when pushed, instantly mutes game sound and picture and transforms the computer-user's screen to a business-like yet non descript "spreadsheet." Additionally, it is estimated that a staggering 50 percent of US workers participate in an NCAA tournament pool—activity facilitated by CBS's online brackets and free group "fantasy" tournament registration.[52]

CBS's "integration" strategy for the NCAA represents a careful matching of sponsors with the sport league's "brand" identity as an amplification of the network's identity across every possible platform throughout the media corporation. For CBS, the relative youth and "early adopter" status of many of those viewers interested in the NCAA tournament's early rounds (not to mention year-round NCAA sport coverage) encouraged the network to focus on its free, traditional broadcast and online digital platforms to reach those viewers, literally, where they live (via television, but also via home and office computing and through the mobile media that individuals carry with them wherever they go). Between 1995 and 1997, just prior to obtaining the NCAA package, CBS acquired a percentage of SportsLine.com, an online portal for sports news and information and for fantasy gaming. With CBS's majority interest, the site was renamed, CBS.SportsLine.com and went public in 1997. Through this site, CBS operated NFL.com for five years (until the NFL moved to consolidate all of its online operations in house, in 2006). In 2007, CBS SportsLine was renamed CBSSports.com "in order to present a clear branding connection between the company's television, online, and mobile interactive sports coverage."[53]

At CBSSports.com, CBS offers headline news and analysis that, beyond the NCAA, focuses on the NFL, NBA, MLB, NHL, college sports broadly, professional golf, tennis, and auto racing. However, as part of the NCAA deal, CBS is the publisher of the official sports site of the NCAA and of the online and cable television outlet CSTV (College Sports Television), which serves as a "homepage" for affiliate campus athletic departments, through which the campus's teams stream video and participate in national contests and polls. CSTV-affiliated programs and sites also function as a portal to merchandising sales venues that are partnered with CBS retail sales via CBSSportsStore.com. CBSSports.com also hosts fantasy sports leagues and regularly features "interactive" contests, for example, a "cheerleader of the week" poll related to NCAA content.

Figure 6.2 This image illustrates the range of college sport information available through CBS's College Sports Network online portal.

Source: http://www.sportsline.com/cbscollegesports/ (accessed October 15, 2008).

When not in thrall to March Madness, CBSSports.com offers streaming video of other sports, such as free, live, online coverage of the PGA Masters golf tournament (originating in 2008). Additionally, CBS partnered with Pontiac to sponsor the CBS Sports NCAA Tournament Channel on YouTube. Thus, all uploads and links on YouTube are solely sponsored by Pontiac and, ideally, then, the sporty GM line is, by association, affiliated with the NCAA competitors' youth and energy, as well as with the "all-access," self-authored appeals of YouTube as the top site for individually uploaded video clips.[54]

Indeed, while sport television is often experienced as communal and larger than life, sport content is also exceptionally well suited to post-network-era, multi-platform and flexible microcast applications. CBSSports.com's streams allow viewers with broadband computer access to watch game coverage from each US region, regardless of the user's actual location. Such access offers an enhancement of the promise of televised sport (offering to "take you there") while the new media application overrides television-specific market constraints that otherwise dictate what games are available to certain viewers. From discrete news alerts or compressed video highlight clips to fantasy league gaming and online outlets, even the most "traditional" broadcast interests now actively integrate over-air and digital, HD and mobile, online and wireless sport advertising and content investments into a coherent, experiential stream.

While watching sport television *on* television, the fan can feel part of a broad community, and while online and with mobile technology in hand she or he is simultaneously addressed as an *individual* with specifically tailored à la carte requests. To borrow a description of NBC Universal's strategy for reaching a target audience of women by combining content from *Today*, Oxygen Media, Bravo Media, and iVillage, the goal is to

> bring the entirety of the NBC Universal portfolio of assets to the advertiser in one package. ... So we hear all the particular attributes and lifestyles and mindsets and brand objectives, and then we come back here and we do a ton of research on our content across everyplace, and so we can actually put together sort of a virtual network just for them.[55]

The truest success for CBS and its partners in content production and advertising, however, is when the individual becomes part of the stream her- or himself, via active participation and "cross-pollination" across platforms.

When individual users register at CBSSports.com, they can "personalize" their site experience (e.g., by selecting favorite leagues, teams or conferences about whom to be updated or alerted). Each such personalization, combined with registration/user biographic data and amplified with further site contributions (e.g., from message-board posts to fantasy league participation), offers a data-mining stream that *both* further individualizes the user's experience (e.g., prioritizing certain photos or headlines when the user logs on) *and* provides increasingly specific market-data and sponsor-user matching techniques that profit CBS. In this sense, the CBSSports.com visitor both participates actively in a vicarious, content-driven, "value-added" à la carte sport "experience" *and* closes the circuit between her- or himself and the market. The user is both a market research subject and an active consumer.

Challenges Facing Sport Programming in the Post-Network Era

As the contours of the post-network era develop, sport programming presents a unique paradox. It remains the most visible, ritual site of "broadcast" address, gathering the largest, multi-demographic audiences for shared, "water-cooler" television experiences. It continues to drive television-specific technologies and their adoption, reflecting and supporting the ongoing strength of television itself and its formal, textual innovations (especially those seen in consumers' sport-driven HDTV adoption and in hi-def and mobile photographic technologies that capture sport events themselves). Simultaneously, however, sport is profoundly "localized" and individualized via Internet and mobile technology-based applications that are fully owned and operated extensions of network brands.

It is important to recall that each transitional period in US television history has been framed by similar logics of "opposition." For example, in the 1970s, cable television was hailed, in contrast to broadcast television, as a democratizing and even revolutionary "new media" force—a technology that could be responsive to particularized and previously underserved audiences in ways that broadcast media could or would not be. While broadcast media were acknowledged as innately national, with programming shared simultaneously by a *mass* audience in exchange for the price of a receiver, cable was trumpeted as an explicitly "narrowcast" medium, responsive to niche audiences in exchange for a subscription fee. Broadcasters battled the increasing competitive threat of cable by claiming to serve a broad public in an era of otherwise waning social cohesion and public debate. The cable industry alternately critiqued broadcasters' claims by arguing that the "mass" logic of broadcasting diluted the force of its content. These competing tensions and claims regarding public service, democratic access, and shared versus niche culture continue to frame industry and public debates about "old" and "new" media, network television, and its post-network extensions. What remains, then, is to think about the complex ways in which "new" media applications are sites of convergence between "old" and new, and to understand television's continued presence and significance in everyday life. Indeed, the post-network era is significantly built on business practices dating from earliest broadcast history.[56]

In 2006, Les Moonves, president and chief executive officer of CBS Corp., commented specifically on the dual network and post-network-era productivity of sport as a site of tradition *within* change. Moonves argued that CBS was the "most stable" network in television, particularly as evidenced by the broad audience attracted to his network's NFL and NCAA Men's college basketball tournament coverage, both of whose online and mobile media presences had taken off for CBSSports.com.[57] Indeed, sport coverage allows the Big Four broadcast networks to position themselves cannily, in Patricia Aufderheide's words, as "the video medium of the poor, the immigrant, the uncabled,"[58] while *simultaneously* cultivating new fan bases (and cross-conglomerate streams) of consumers who are among the most tech-savvy and reflective of the most valuable "niches" among all media users.

It might be productive at this historical juncture to consider how sports leagues and conferences themselves have become fully fledged media institutions of the post-network era. In this chapter's examples, for instance, the NFL and NCAA exert content controls on coverage of their events and licensing of the images therein. The league and conference also reserve right of approval regarding televised, online, and mobile media sponsors and partners. Both entities may exert influence upon and approval of on-air talent as well. With such practices, sport institutions are, arguably, revising the historic role of the broadcast-era television sponsor for a new media, post-network era. For example, in radio and television through the 1950s, the sponsor packaged every aspect of the program for the network to distribute to the audience, including all "story" content, on-air talent, and advertising. The updated twist here, perhaps, is that, in addition to the network and its sponsors, the sport institution now profits from the market data and consumption of the viewer/user as well.

Thus, while there is admitted excitement and pleasure in participation, access, and interactivity, we might also need to consider how individual engagement of sites such as CBSSports.com, the *reach* of its mobile alerts, and the site's tracking of our sporting passions fundamentally serves the interests of the sport media's producers and advertisers. How might the "promise of interactivity" also represent labor within and for the market? That is, how does sport television and its extensions turn "passion" into profit, and for whom?[59] Arguably, such new-media imperatives profit most from further atomization of audiences or from narrower and narrower niche interests being carved out of the broader sporting community. We might ask what this means for our sense of engagement with community, broadly conceived, especially when not all sport fans have access to such technologies or modes of "personalized," flexible microcasting.

While much popular, trade-industry, and academic rhetoric is currently engaged in promoting the idea of network death, attention to the popularity and broad audience appeal of sport media suggests a much more nuanced post-network multi-media platform reality. Arguably, sport and network television are particularly symbiotic US cultural institutions. They appear uniquely apolitical *and* are simultaneously our most visible indicators of whom and what are most valued within contemporary US culture (as seen in sponsor dollars, "ideal" target audiences, scheduling practices and the featured sports themselves). Such characteristics of sport and television might lead us to consider, for example, with the exception of the Olympics and certain play-off schedules, why, in such a proliferate sport television universe, there is no viable "market" for women's professional soccer, or no national broadcast rights deal for the National Hockey League. Or why, though Internet and mobile media applications are relatively "new" media, their *use* has typically replicated broadcast television's rather traditional modes of conceptualizing "audience," or "target", markets. As a recent feature in *Advertising Age* noted, where Internet and new media "ad dollars are concerned, ... it seems to be shaping up to look a lot more like the broadcast world, with a handful of players dominating marketers' spending. TV's Big Four environment has turned into the Big Four online."[60]

While changes in business practices and the competitive media environment have undeniably altered the nature of television's cultural significance from the zenith of the network era, continued engagement with "traditional" television by the broad US public and the complementary effect of new media technologies upon television viewership still insistently point to the medium's past, present, and future significance as a shared site of cultural production and to the apparent, lingering need for television in these terms. It remains important to consider, in the face of contemporary competition from all media corners, the calculated rhetorical value to the traditional networks' self-portrayal as inherently "democratic," "public service" media. However, while much recent scholarly and industrial rhetoric has speculated on the loss of a sense of place, more than any other mode of television, sport broadcasting continues to engage broadly and to construct community *as* broadly communal *and* specifically place-based in ways both powerfully shared and individually experienced.

Notes

1 Chris Pursell, "Sports: TV's Power Play," *Television Week*, 25 August 2008, 1; http://www.tvweek.com/news/2008/08/sports_tvs_power_play.php.

2 Ibid., 26.

3 Ibid.

4 Ibid.

5 Horace Newcomb and Paul M. Hirsch, "Television As a Cultural Forum," *Television: The Critical View*, ed. Horace Newcomb (4th ed., New York: Oxford University Press, 1987), 505. These points about sport as cultural forum were also made in a presentation by the author, "Sporting Community? TV as Cultural Center and 'Level' Playing Field?," at the Flow Conference, Austin, TX, October 2006, accessible at http://www.flowconference.org/rt22johnson.doc.

6 Newcomb and Hirsch, ibid., 506. I would argue that televised sport programming, more than any other genre, actively engages and encourages explicit discussions of community identity and ideals. Sport TV references race, gender, age, sexuality, and regional "difference" in ways that appear "safely" displaced onto "games" and "play," but that, effectively, represent the shared ritual site for significant interrogation of these culturally constructed categories.

7 Ibid.

8 Henry Jenkins, *Convergence Culture: Where Old and New Media Collide* (New York: New York University Press, 2006), 2.

9 John B. Horrigan, "Home Broadband Adoption 2008," *Pew Internet & American Life Project* July 2008, http://www.pewinternet.org/ppf/r/257/report_display.asp, accessed August 27, 2008. According to the Pew report, 55 percent of Americans now have broadband connections at home, though growth in broadband adoption remains flat among the poor and among African Americans (p. v).

10 Matthew Futterman, "Olympics Spark Contest Among TV Broadcasters," *Wall Street Journal*, 25 September 2008, http://online.wsj.com, accessed September 26, 2008.

11 Andrew Krukowski, "Olympics Seeking Viewership Record," *Television Week*, 25 August 2008, 3; http://www.tvweek.com/news/2008/08/olympics_seeking_viewership_re1.php.

12 Robert W. McChesney, *The Political Economy of Media: Enduring Issues, Emerging Dilemmas* (New York: Monthly Review Press, 2008), 213.

13 Ibid., 219, 221.

14 Michael J. Socolow, "'Always in Friendly Competition': NBC and CBS in the First Decade of National Broadcasting," *NBC: America's Network*, ed. Michele Hilmes (Berkeley: University of California Press, 2007), 36.

15 Michele Hilmes, *Hollywood and Broadcasting: From Radio to Cable* (Champaign: University of Illinois Press, 1990).

16 For prime-time television schedules from the medium's inception to the present, see Alex McNeil, *Total Television* (4th ed., New York: Penguin, 1996).

17 Michael Oriard, *Brand NFL: Making and Selling America's Favorite Sport* (Chapel Hill: University of North Carolina Press, 2007), 12.

18 Though, interestingly, from 1953 to its demise in 1955 the much overlooked DuMont network carried NFL games on Saturday nights. Starting with the 1956 season, NFL games were carried by CBS.

19 The proprietary nature of NFL telecasts has since shifted. CBS carries AFC games and FOX carries NFC games. NBC, ESPN, and the NFL network offer interleague and alternating conference coverage.

20 Gary R. Edgerton and Jeffrey P. Jones, "Introduction: A Brief History of HBO," *The Essential HBO Reader* (Lexington: University Press of Kentucky, 2008), 1.

21 Ibid., 2.

22 "DISH Network Launches WGN America in HD and Renews Transmission Agreement with Tribune's 23 Television Stations," http://news.cnet.com, 12 June 2008, accessed August 8, 2008.

23 In 1995 the Walt Disney Company purchased Capital Cities/ABC and thus became full owner of ESPN.

24 ESPN began its diversification and expansion in the early 1990s with the launch of its international networks (from 1989 to the present); its merchandising and "lifestyle experience" division, ESPN Enterprises; ESPN Radio (1992); nine (and counting) domestic/US cable outlets; ESPN on ABC, which subsumes ABC Sports; ESPN Online; ESPN Interactive; ESPN Wireless; ESPN Popular Music, Recording and Distribution; ESPN Sports Zone(s); ESPN X-Games; ESPN Winter X-Games; World Series of Poker; Arena Football (with Russell Athletics); and ESPN Visa (with Washington Mutual Bank). Notably, in 2006, *Monday Night Football* moved from ABC to its Disney Co. sibling. ABC's sports division has, since, effectively been "offshored" to ESPN. College football games on ABC are produced by ESPN, feature ESPN talent, and are telecast under the banner "ESPN on ABC."

25 FOX began its full seven nights per week program schedule in 1993. However, at present the network still programs a "limited" prime-time schedule only, with original programs airing from 8:00 to 10:00 p.m. eastern/Pacific, rather than from 8:00 to 11:00 p.m.

26 From this period to the present, News Corp. has been a media conglomerate characterized by vertical and horizontal integration, with, among media properties, ownership of newspapers,

magazines, film studios and libraries, recording labels, broadcast television outlets, cable television outlets, and DirecTV satellite service. It has been a global entity from its inception, with holdings across Europe, the United States, and Australia, particularly.

27 Jim McConville, "Fox Heats up Cable Sports Competition," *Electronic Media*, 17, 17 August 1998, 2.

28 Allison J. Waldman, "NFL Net Cries Foul at MSOs," *Television Week*, 13 November 2006, 14.

29 Oriard, *Brand NFL*, 3.

30 Ibid., 174.

31 Ibid., 175. Referring here, specifically, to the 2005–6 NFL season. NBC's *Football Night in America*, new to Sunday prime time in 2006, has additionally been hailed as a "fiscal touchdown" with tremendous spillover effect for NBC prime time across the week schedule: "Without sports, NBC's rating among viewers ages 18 to 49 is up 4 percent. Counting sports it's up about 15 percent." Jon Lafayette, "Ebersol's Gridiron Groove," *Television Week*, 9 October 2006, 26.

32 While we now think of HDTV and sport as linked, it is also important to think historically about the ways in which sport on TV was used to promote the original adoption of television and initial purchases of television receivers for the home. Additionally, sport TV is often considered to have been a driver for the adoption of color receivers, and more recently it has driven adoption of satellite services such as DirecTV (home of the NFL's subscription-based "Sunday Ticket" package). It is productive, in other words, to think about a much longer history of collaboration and connection between sport and the promotion and adoption of TV technology.

33 Chris Pursell, "Sports Programming Surges," *Television Week*, 3 March 2008 1; http://www.tvweek.com/news/2008/03/sports_programming_surges.php.

34 Scott Collins and Lynn Smith, "NBC, Like Phelps, in Record Pursuit at Olympics," *Los Angeles Times*, 13 August 2008, A15.

35 As an example of wreaking havoc on televisual flow, the 2008 Major League Baseball All-Star Game at Yankee Stadium took FOX by surprise. Pre-game festivities began at 8:00 p.m. eastern time, with the first pitch thrown just after 8:45 p.m. However, the game lasted an unprecedented fifteen innings. The final out did not occur until almost 2:00 a.m. eastern time. Time zones play a significant role in sport TV that is *not* generally a factor in other regularly scheduled programming (though a common issue for "event" programs such as awards shows). For example, the "early afternoon" NFL game starts at 10:00 a.m. on the West Coast. However, in spite of this temporal uniqueness of sport TV within broadcast flow, overall, the business of sport and of TV collaborate to assure that the "9-to-5" laborer-viewer is still positioned as the primary target audience, with weekend sport programming and prime-time weekday sport encouraging "lucrative" viewer attention and restoration for the work/marketplace the next week/day.

36 Margaret Morse, "Sport on Television: Replay and Display," *Regarding Television: Critical Approaches—An Anthology*, ed. E. Ann Kaplan (Frederick, MD: University Publications of America/American Film Institute, 1983), 48.

37 Ibid., 60.

38 Steve Rushin, "The Titan of Television," *Sports Illustrated*, 81, 16 August 1994, 36. As regards the news and spectacle hybridity of sport programming, it should be noted that Arledge also developed *Wide World of Sports* (ABC, 1961–98) and *The American Sportsman* (ABC, 1964–81). He was president first of ABC News and Sports (1985–90) and then of ABC News (1990–98), where he developed the prime-time news magazine series *20/20* (ABC, 1978–) and *Primetime* (ABC, 1989–). Arledge's protégé, Dick Ebersol, was, similarly, president of NBC Sports and then senior vice president of NBC News before assuming his current position as president of NBC Universal sports and Olympics.

39 Morse, "Sport on Television," 50. It should be noted here that, when "bells and whistles" are absent from sport TV, telecasts often fail. NBC discovered this in 1981, with an ill-fated attempt to broadcast an NFL pre-season game with ambient sound only. Without color and

play-by-play to "anchor" the enhanced image technology aurally, viewers felt cheated and adrift. This is a good example of the difficult balance sport TV must strike between "realism" and spectacle. Ambient sound pushed the balance toward "too real" while, simultaneously, seeming entirely stilted and "false."

40 "Clicks: Technology We Have, Technology We Will Have," *Television Week*, 11 August 2008, 6.

41 Natalie Finn, "Sports Is the Driver Behind HD Revolution," *Television Week*, 30 October 2006, 25.

42 Meg James, "NBC Adds a Gal Pal to its TV Holdings," *Los Angeles Times*, 10 October 2007, C7.

43 As Christopher Anderson has argued, sporting events such as the Olympics, in particular, "still have a unique aura, a value that cannot be duplicated and has not yet been diminished in the multichannel television universe." See Anderson, "Creating the Twenty-First-Century Television Network: NBC in the Age of Media Conglomerates," *NBC: America's Network*, ed. Michele Hilmes (Berkeley: University of California Press, 2007), 283. And Collins and Smith, "NBC, Like Phelps," A15.

44 Collins and Smith, ibid., A1, and quoting NBC research chief Alan Wurtzel as regards Internet use at NBC.com and TV viewing on NBC, MSNBC, and USA during the Beijing Summer Olympics, 2008. See also Douglas Gomery, "The Coming of Television and the 'Lost' Motion Picture Audience," *Journal of Film and Video*, 38,1 (1985), 179–85, for a historical interpretation of television's introduction and the "lost" film audience, which posits TV, its "free" programming, suburban development, and the baby boom as "complementary goods" in relation to film's waning audience in the 1950s.

45 Lisa Parks, "Flexible Microcasting: Gender, Generation, and Television–Internet Convergence," *Television After TV: Essays on a Medium in Transition*, ed. Lynn Spigel and Jan Olsson (Durham, NC: Duke University Press, 2004), 134. Emphasis mine.

46 Targeting, for example "brand in hand" marketing via cell phones and other hand-held personal communication devices that can deliver up advertising (brand) messages directly to individuals "in hand" or on their person, via those technologies. See Fareena Sultan and Andrew Rohm, "The Coming Era of 'Brand in the Hand' Marketing," *MIT Sloan Management Review*, Fall 2005, 83–90.

47 Ibid., 85.

48 Richard Sandomir, "CBS to Acquire CSTV for $325 Million in Stock," *New York Times*, 4 November 2005, C3.

49 Chris Pursell, "CBS's March Madness," *Television Week*, 19 March 2007, 1.

50 Ibid.

51 Chris O'Brien, "NCAA is Tough on Workplace Networks: Watching Video Can Disrupt Companies' Computer Systems," *Lexington Herald-Leader*, 21 March 2008, B6.

52 Andrew Eder, "The Office Pool," *News Journal*, 24 March 2008, F1.

53 "CBS Sports.com," *Hoovers Company Records*, 8 July 2008.

54 Beyond such overtly corporate spaces within YouTube, the site is phenomenally popular for ripped sports highlights, lowlights, trash talk, school pride and self-promotion.

55 Jon Lafayette, "Cobbling Together a Target Ad Audience," *Television Week*, 5 May 2008, 22; http://www.tvweek.com/news/2008/05/cobbling_together_a_target_ad.php. Here, Lafayette is quoting NBC Universal senior vice president/head of marketing, Debbie Reichig.

56 Please see a further elaboration of this argument in Victoria E. Johnson, "Historicizing TV Networking: Broadcasting, Cable, and the Case of ESPN," *Media Industries*, ed. Jennifer Holt and Alisa Perren (Malden, MA: Wiley-Blackwell, 2009).

57 Joe Mandese, "Beyond the Top 25-Markets," *Television Week*, 26 January 2004, 10.

58 Patricia Aufderheide, *Communications Policy and the Public Interest: The Telecommunications Act of 1996* (New York: Guilford Press, 1999), 48.

59 Mark Andrejevic, "iMedia: The Case of Interactive TV," *iSpy: Surveillance and Power in the Interactive Era* (Lawrence: University Press of Kansas, 2007), 135–60.

60 Abbey Klaassen, "The Short Tail," *Advertising Age*, 27 November 2006, 1.

Chapter 7

A Form in Peril?

The Evolution of the Made-for-Television Movie

Erin Copple Smith

Since its heyday in the 1970s and 1980s, the made-for-television movie has become as much punchline as television form. For years, scholars and audiences alike have disparaged the outrageous, "ripped from the headlines," melodramatic stories the form relied upon in the 1980s and 1990s. Titles such as *Mother, May I Sleep With Danger?*, *A Face to Kill For*, and even *A Face to Die For* have become the source of countless jokes, and cable theme nights such as Lifetime's "Killer Babysitter Evening" highlight the campy nature of the form. Nonetheless, there can be no mistaking the historical importance of the made-for-TV movie within the television industry; the form was once a central narrative form for television storytelling. Indeed, in his book *Inside Prime Time*, Todd Gitlin notes that "The three networks now underwrite more original movies than all the studios combined" and that the films composed a full 25 percent of prime-time programming slots at the time.[1]

Despite their proliferation until the mid-1980s, made-for-TV movies have been slowly but steadily disappearing from broadcast airwaves.[2] Although the form was once as central to broadcasters as series, the advent of the multi-channel transition introduced particular challenges for it and created new opportunities that significantly altered its industrial and textual aspects. Investigating the history and current state of the made-for-TV movie consequently requires looking beyond broadcast schedules and assessing a range of institutional and textual factors. The arrival of cable and the related fragmentation of the broadcast audience caused economic changes that led to the migration of the form from broadcast networks to cable channels. It has since developed a successful and abundant existence on cable channels, for which its narrative adaptability has proven to be particularly useful in establishing distinct channel identities or brands. A range of basic economic features, industrial imperatives, and institutional norms give cable networks the advantage over their broadcast competitors as distributors of made-for-TV-movies. Additionally, shifts in the global television market, including trade quotas, growth in foreign markets, and the rise of international co-production, have all affected the production of made-for-TV movies, the types of stories they tell, and the markets that they reach. This chapter examines each of these elements in order to demonstrate the continued significance of the form within the contemporary television industry. Unlike the programming forms that weathered the multi-channel transition with slight and steady adjustment, various textual and industrial shifts led to radical changes in the made-for-TV movie well before the post-network era began further reconfiguring the television industry.

Originally developed as a means for broadcast networks to air special event programming that could be produced relatively inexpensively, the made-for-TV movie enjoyed a long life as a staple of broadcast line-ups. Although the term for the form within the industry itself is "movie of the week," or "MOW," this terminology proves to be less precise than the term used here. The MOW terminology dates to the form's origins and the networks' scheduling strategy of holding a place for it on weekly schedules, as in the case of the *CBS Saturday Movie*, for example. These regularly scheduled made-for-TV movie nights began disappearing in 1976, at which point networks scheduled a mix of made-for-TV movies and theatricals in the remaining schedule blocks devoted to films. The term "made-for-TV movie" addresses both the industrial and textual features of the form regardless of scheduling.

Although scholars and audiences alike seem to have forgotten about the form's long centrality to broadcasting, the made-for-TV movie successfully adapted to the multi-channel transition and, far from dead, came to thrive on cable during the post-network era. Both the form and its evolution reveal key aspects of changes occurring within the television industry over a span of almost 50 years. Rather than relegating it to the dusty pages of television history, pulling it into the spotlight provides essential insight into many of the key industrial transformations of television over the last half-century.

The Made-for-TV Movie in the Network Era

Broadcast television networks have always aired movies, but before 1964 they typically did not produce or license movies created specifically for a first run on their networks.[3] Rather, the most common form of movie on television schedules until the 1960s was the "theatrical"—the industry's term for "Hollywood" films first released for screenings in theaters. But as the cost of licensing theatricals from studios grew steadily, the networks sought alternative ways to schedule films that would allow them to maintain desired profit margins. The made-for-TV movie developed to solve the problem of their reliance on the popular theatricals, which became a sizable component of their broadcasting schedules. The networks began decreasing their reliance on theatricals in the late 1960s, at which point the cost of acquiring movies from Hollywood had grown so steep that they decided to begin licensing films made specifically for them.

Procuring made-for-TV movies from independent distributors proved less expensive than either buying theatricals or making their own films, but allowed them to maintain this significant component of their schedules.[4]

The made-for-TV movie thus originated as a way for the networks to duplicate the ratings success of the theatricals for far less money, although the form's origins can be traced to well before these financial pressures developed, all the way back to early television.[5] Gary Edgerton explains that the move of larger Hollywood film studios into producing television in the late 1950s hurt independent television producers (Ziv-TV, Desilu, etc.) who had been producing television series since the late 1940s. Producers such as Jennings Lang attempted to stave off this Hollywood invasion by innovating new programming forms for television, including the "special event" and "long-form" programs that were the precursors of the made-for-TV movie. These forms were promoted as something beyond the typical half-hour or hour-long series and were attempts on the part of the networks to draw audiences to distinctly televisual offerings. They operated as alternatives to theatricals and were able to draw similar audiences. Lang eventually became a top executive at Universal TV, where he continued to innovate with long-form programming; he ultimately developed a series of "TV epic[s] (or special events)" for which a network devoted "an entire evening … to a single spectacular, made for the occasion."[6] In 1963, Lang and Universal TV convinced NBC to invest in "mini-movies" as an alternative to theatricals. This program form intrigued the network on account of the indisputable popularity of the theatricals and the burden of their rising costs.

Early incarnations of made-for-TV films were quite different from the overwrought, melodramatic material commonly considered to be a hallmark of the form today. These earlier films were much more similar to Hollywood offerings than their successors and featured storylines and casts straight out of Hollywood. This similarity eroded over time as television networks became increasingly interested in developing stories with a stronger social message than typical Hollywood films. Edgerton explains that by the 1970s the form was changing and becoming more socially relevant than ever before, noting that ABC's *Movie of the Week* series was interested in developing "present-day stories" rather than more traditional generic fare such as Westerns or fictional melodramas.[7] This social relevance characterizes many made-for-TV films of the early 1970s through the 1990s and must be understood to have industrial, economic, and cultural significance.

Narratively, television movies have always been able to deal with more socially relevant topics than series television due to their one-time nature. The lack of recurring characters or settings that distinguish television series allowed them to tackle such topics as AIDS (*An Early Frost*, 1985), race (*Brian's Song*, 1971), and issues including domestic violence (*The Burning Bed*, 1984) long before series television took such risks. Douglas Gomery notes that "topical or controversial material not deemed appropriate for regularly scheduled network series" found its home within made-for-TV movies.[8] Likewise, Elayne Rapping explains, "TV movies offer opportunities of a kind of work not easily done elsewhere in television or film and therefore tend to attract from the start producers, actors, and writers with more politicized agendas."[9] Because of its particular place within television programming, Rapping claims, "it is safe to say that in the realm of TV genres, the telefeature is at any given time the most suitable form for dealing thoroughly not only with complex social issues but also … with the most resonant of tensions, contradictions and ambiguities."[10] The form's characteristic "event" or one-time status allowed these themes and agendas to become part of TV schedules without the worry of irreparably offending audiences.

The attention to controversial issues served several industrial purposes. As Laurie Schulze argues, made-for-TV films' inclusion of "controversial issues" contributed to their marketability.[11] She posits that the salacious topics provided a solution to an industrial challenge of the form: how to promote them. Unlike series that are likely to return an audience week after week, made-for-TV films demanded that networks develop significant marketing campaigns, especially after eliminating their weekly schedule position. Schulze contends that this problem of promotion was solved, at least partially, by the films' sensational nature, which functioned as an easy way to attract audiences intrigued by the "issue of the week." Not only did their sensational nature not turn away audiences, it actually drew them in.

The made-for-TV movie exists as a significant part of US television for both cultural and industrial reasons. The disappearance of such films from network television, their relocation to cable channels, and the textual shifts that occurred within the form as the multi-channel transition changed the rules of television production likewise can be attributed to economic factors that yield significant cultural implications as well.

The Made-for-TV Movie in the Multi-Channel Transition

Critics and scholars alike tend to regard the diminished presence of the made-for-TV movie on broadcast networks as evidence of its decline and decreased relevance to television. The truth is that the form has never disappeared completely from either broadcast or cable television. Rather, it slowly faded in popularity among both audiences and industry executives, which led to a gradual reduction in the number made, particularly on broadcast television. ABC's *Movie of the Week*, the last series to air exclusively made-for-TV films, concluded its seven-year run in 1976, and NBC's *Saturday Night at the Movies* ended in 1978 after seventeen years. Even with the loss of these dedicated series, made-for-TV movies continued to air on broadcast television throughout the 1990s and into the early 2000s, albeit much less frequently, and variations persisted into the early 2000s that offered some regular presentation of the form. These included a mix of made-for-TV movies and theatricals, but all regular schedule positions for movies were eliminated at the beginning of the post-network era. *The Wonderful World of Disney* aired on various broadcast networks until finally leaving ABC in 2005, and CBS last offered the *CBS Sunday Night Movie* in the 2005–6 season. In 2006, *Variety* reported there would be no "regularly scheduled movie nights" on broadcast television for the first time since 1961.[12] Likewise, *Television Week* reported in the same year, "For this coming season, CBS has thrown in the towel, as ABC and NBC did before it, and canceled its two-hour Sunday night movie block."[13]

Although these series ceased relatively recently, their decline was gradual. Movie nights remained immensely popular into the mid-1980s, but by the end of that decade—a point by which the adjustments of the multi-channel transition were apparent throughout the industry—networks were beginning to lose faith in the program form. A 1987 *Advertising Age* article reported on the diminished stature of the form: "Movies, which occupied as many as five nights a week in the 1986–87 prime-time TV season, will figure in ratings battles on just three nights this fall."[14] According to reporter James P. Forkan, made-for-TV movies were losing popularity, and ratings were down 9 percent from the previous season.[15] In 1988, NBC's vice president in charge of movies for television, Tony Masucci, told the *New York Times*, "I hope some of these movies will do better than they would have last season. But I'm not holding my breath for gigantic shares. Basically, the ratings have been going down over the last few seasons, and I'm not expecting a dramatic change in that pattern."[16] In the article, Masucci himself points to many of the features of the multi-channel transition as part of the waning popularity of made-for-TV films and mini-series: "Cable has increased tremendously, and home video has taken away part of our audience."[17] By 1989, *USA Today* reported that then NBC president Brandon Tartikoff was considering dropping one of the network's two (and sometimes three) dedicated movie nights and replacing it with a regular series in response to complaints about the poor quality of the films.[18] Tartikoff, defending the network against critics who were unimpressed with the network's original films that season, explained, "If I get a really good [series], I'll knock out a movie and, right there, half of this so-called problem that may exist in our TV movies will disappear," indicating his lack of commitment to a form that was apparently easier to replace than rehabilitate.[19] Essentially, it was becoming clear by the late 1980s that ratings for special event programming such as the made-for-TV movie were down considerably from the form's height of popularity in the 1970s. Audiences wanting to watch movies had options far beyond the offerings of broadcast networks; they could turn to cable for a wider selection of both theatricals and made-for-TV movies or they could rent a home video to watch on their own timetables. Industry executives were thus left scrambling, trying to restructure their schedules to cut down on the number of films they were airing and increase the exposure of those they chose to air.

Despite the diminished presence of the made-for-TV movie on broadcast networks, which has led so many to claim the death of the form entirely, various developments in the multi-channel transition have allowed the form to remain very much alive—although with some changes. First, the gradual distribution of cable during the 1980s led to countless changes within television and created a new outlet for made-for-TV films. Although broadcast networks were increasingly reluctant to invest in films, given their decreasing audiences and the challenges of promotion, the form fit well with the economic and competitive needs of the burgeoning cable field. Secondly, the growing internationalization of the television industry provided some challenges for the melodramatic, social issue-based version of the made-for-TV film that had become standard on broadcast networks. Yet, after some textual negotiations, expanding international co-production practices have also led to another new installment in its ever-evolving history.

The Made-For-TV Movie on Cable: Industrial Advantages, Network Branding, and Prestige Programming

A 2001 *Daily Variety* article reported:

> It's estimated that 15 basic cable nets will air nearly 100 movies over the next year, up drastically from 1996, when just four nets premed 54. On the broadcast side, ABC, NBC, and CBS will air roughly 45 original pics this season, down from last season's 59, which had declined from the previous year's 74.[20]

Why did the made-for-TV movie thrive on cable as it fell from favor among broadcasters? Even though we might often generally conflate broadcast and cable as constituents of a common television industry, their remarkably distinctive institutional structures explain the different status of the made-for-TV movie on each outlet after the network era.

Cable channels are supported by a different revenue structure than broadcasters, which enables the decreasing audience sizes and increasing costs of the made-for-TV movies to be a more profitable enterprise for cable. Cable channels are also less encumbered by the traditional scheduling practices that limited broadcasters and had decreased the viability of the form in its previous home. Finally, a number of the attributes of the made-for-TV movie correspond well with the programming needs of cable channels seeking to establish a particular identity and to break out of the vast glut of cable channels lacking word of mouth or an original appeal. The films can be imprinted with elements designed to echo any cable network's brand identity, be it ESPN's sports-themed films or the female-targeted films of Lifetime.

Together, cable and made-for-TV movies often proved an ideal pairing, though understanding the success of the latter on cable channels cannot be achieved by considering economic and industrial factors alone. Their value derives from more than simply considering the cost of programming: value involves indirect benefits, such as further establishing the brand of a channel or conferring prestige on its endeavors in addition to weighing production budgets and audience ratings. The cheapest option for cable channels is often to buy a program that has already aired on a broadcast network, but scheduling such programming does little to establish a distinctive identity for the channel, thereby limiting its value. In addition to being a prudent economic choice, made-for-TV movies also provide considerable value in terms of how they can be used in the task of channel branding and in attributing prestige to a channel.

Cable's Industrial Advantages

The particular economics of cable provide a structuring advantage relative to broadcast. Cable networks derive their revenue partially from advertising sales, but also through the subscription fees paid by the cable services in exchange for carrying each channel on their service. This second revenue stream leads to a different calculus of success and enables programs with much smaller ratings than those on broadcast networks to be considered hits. In a 1999 article, *Variety* writer John Dempsey described the appeal of the made-for-TV movies to cable channels: "One reason for cable's rush to make movies and minis is that, quite simply, lots of people watch them."[21] Dempsey, who had reported two years earlier that audiences for the made-for-TV movie were dwindling on broadcast television, does not contradict himself, because "lots of people" on cable still amounts to a poor showing for broadcast television. Although the made-for-TV movie was no longer successfully attracting a 30 to 50 share—or 30 to 50 percent of those watching television at that time—for the broadcast networks, the smaller audience it was able to draw on cable was more than adequate for those channels. The Hallmark channel, for example, reportedly doubles its average prime-time ratings when it elects to air an original movie.[22] Dempsey also reported that the channel's *The Good Witch*, which aired in January 2008, scored "an *impressive* 3.8 household rating."[23] To put that rating in context, NBC's February 2008 presentation of *Knight Rider* as a made-for-TV movie was viewed as successful for broadcast standards, with a 5.0 household rating—the highest rating scored by a broadcast made-for-TV movie since *Their Eyes Were Watching God* in 2005.[24] Notably, although "successful" for a broadcast made-for-TV movie, *Knight Rider* was beaten in its final hour by the ABC series *Extreme Makeover: Home*

Edition.[25] These figures suggest that made-for-TV films airing on broadcast networks very rarely fare any better than their regular series, and often fare much worse. In contrast, cable networks such as Hallmark and Lifetime often boast their highest viewership when airing original movies, although even these ratings rarely come close to the viewership required to be a moderate success on broadcast.

This different calculus for success on cable channels is augmented by the thematic fit possible among films and the specific and niche appeal that many channels have cultivated. Cable channels can survive without the traditional broadcast strategy of trying to gather a mass audience, and this eases the challenge of developing films, because whatever common interest distinguishes the channel is likely to be shared among the established audience. This creates a set of themes or topics likely to be successful that is more specific than the general salaciousness Schulze describes as central to the promotion of the films on broadcast networks.[26] Instead of trying to appeal to audiences through sensational themes, cable channels draw viewers to movies by targeting precise interests—a strategy unavailable to broadcasters.

The dual revenue stream does advantage cable channels in some ways, but because their audiences are in most cases much smaller than those of the broadcasters they often also command significantly reduced rates from advertisers. Cable channels are consequently very interested in affordable yet distinctive programming, which made-for-TV films can provide. As cable channels were on the rise throughout the 1980s and 1990s, they struggled to fill their schedules with available budgets. Original scripted programming often draws a larger audience than the syndicated programs that compose so much of cable schedules (shows such as *Law & Order* that previously aired on broadcast networks) and they also provide an opportunity to communicate the channel's brand clearly. However, developing original scripted programming to fill an entire week is essentially cost-prohibitive and involves the considerable risk of series production. In comparison with a series, made-for-TV movies are a relatively inexpensive form of original programming that provides an elegant solution to the obstacles facing new cable channels. *Variety*'s Dempsey explains that in 2004 the production costs for films for Lifetime, Hallmark, and Sci-Fi was "bargain basement—$2 million tops for 30 or so of the Lifetime titles and just about all of the Sci-Fi movies, and $1.8 million for the Hallmark ones."[27] In comparison, Dempsey reports broadcast series budgets in the same year were increasingly reaching $3 million per hour for dramas. Contrast those figures with reports from as recently as 2007 that dramatic cable series budgets are only about 75 percent of their broadcast counterparts, ranging from $1.5 to $2.5 million per hour.[28] Even as production costs for original programming have risen, cable channels have continued to pay considerably less for original fare than broadcast networks.

Additional economic worth also resulted from the fact that cable channels were able to gain considerable value from their films because of viewers' different expectations of cable schedules. Cable channels fill their schedules by repeating their programming multiple times a day or week. A 2004 *Daily Variety* article notes, "Original movies also make more business sense for cable than for broadcasters. They are a significant financial investment and broadcast nets only have one opportunity to air them."[29] For example, in 2003, *Blessings*, a *CBS Sunday Movie*, drew 8.2 million viewers in its single scheduled airing.[30] Four months later, TNT garnered a total of 23 million viewers over the opening weekend of its Neil Simon adaptation *The Goodbye Girl* but achieved this figure by cumulating the audience and reshowing the film at 8:00 on Friday, Saturday, and Sunday nights.[31] Cable channels are able to secure considerable programming hours and continual revenue out of their initial programming investment through repeated airings in a manner unavailable to broadcasters. Additionally, in cases in which the cable channels owned a stake in the films, they could also earn profits though subsequent sales of the films to other exhibitors—typically networks in other countries. These other revenue streams also could encourage higher spending. Dempsey paraphrases Steve Koonin, executive vice-president of TNT and TBS, who explains, "The richer the production values, the longer the shelf life. Bigger-budgeted movies also stand a better chance of earning lots more money from markets outside the US"[32]

Channel Branding and Prestige Programming

But regardless of the considerable economic and industrial advantages for cable channels, the made-for-TV movie arguably delivers equivalent—if not greater—value as an opportunity for channel branding or establishing prestige. McMurria explains that, following the explosion of cable channels in the late 1980s, the most successful channels were often the ones that had best branded themselves with a particular identity.[33] To establish these identities, many networks—including TNT, USA, A&E, Lifetime, VH1, E! Entertainment, BET, and the Sci-Fi Channel—used made-for-TV movies and mini-series. Fledgling cable networks were inclined to develop an identity around a particular genre, as in the case of Sci-Fi, and then relied "on high-profile original movies and mini-series to promote and distinguish network identities."[34]

Lifetime, a network that branded itself from the mid-1990s through the mid-2000s as "Television for Women," made particular use of the made-for-TV movie in its schedule. It grew infamous for featuring sensationalized, melodramatic original films, which Eileen Meehan and Jackie Byars identify as the most successful programming strategy in the channel's first fifteen years.[35] The network's films contributed significantly to Lifetime's success, in terms of both drawing an audience and establishing a clear brand. Recently, the network has attempted to distance itself from the more sensational stories that made its original movies famous.[36] It now prefers to develop and air films more directly related to contemporary social issues, particularly favoring the stories of "real women." Amanda D. Lotz explains, "'Issues' are also central to Lifetime's identity—an attribute identifiable in the made-for-Lifetime films."[37] Thus, more recent offerings from Lifetime include *Why I Wore Lipstick to my Masectomy* (based on the experiences of a woman who wrote a memoir by the same name) and *A Girl Like Me: The Gwen Araujo Story* (which deals with transgender issues). Even as the channel has attempted to distance itself from the overwrought melodrama associated with the films and the older, middle-American viewer perceived to watch, it has never given up the form, but has shifted the themes and developed series with different attributes. Indeed, even as late as 2003, Lifetime aired four to five original movies per day.[38] Just like its more salacious early offerings, these films that attend to "real women" and the issues they face reflect the network's attempt to coordinate their original movies with their brand.

Of course, over time, cable networks have increasingly used the made-for-TV movie to advance competitive strategies in conjunction with branding. The form offered cable networks an opportunity to develop branded prestige programming with far less financial risk than developing original series, the other common strategy for delimiting a brand and soliciting prestige. Although cable networks often build their initial schedules based on a combination of syndicated material and unscripted programming (both of which are much less expensive than original scripted programming), many networks found made-for-TV movies to be a logical bridge between these inexpensive forms and the high cost of producing original scripted series. But cost-management is not always the primary consideration, and networks such as TNT and A&E have chosen to invest in high-budget, high-profile made-for-TV films designed to draw the attention of critics. Expensive or otherwise exceptional films provide cable channels with an opportunity to draw critical attention and even awards that are normally monopolized by broadcast products. Such coverage also functions as a form of promotion. In the 1999–2000 season, McMurria reports that TNT produced films with budgets averaging $8 million, and going as high as its $24 million adaptation of *Animal Farm*.[39] McMurria offers this telling quotation from Robert DeBitetto, vice president of TNT Originals: "These are such a part of our branding architecture, that if a movie gets Emmy nominations and critical raves and gets us attention, then that's a home run for us."[40] Similarly, *Television Week*'s Elizabeth Jensen quotes TNT's senior vice president of original programming, Michael Wright, who thinks of made-for-TV movies as "lighthouses" that encourage viewers to "notice you in a way they might not otherwise. It communicates quality to people; your care and enthusiasm for it says so much about the network and what viewers can expect from the other programming. It becomes a calling card."[41] In another article, Wright takes a similar tack, explaining, "We look to our original movies to expand our reach, to make people aware of us who might otherwise not have tuned in. An original movie shines a light on our brand and says, 'This is what TNT is about.'"[42] In this way, the form offers simultaneously both branding and the critical appeal of prestige programming.

The potential of made-for-TV movies as "stunt programming" has proven very attractive to cable executives. A 2004 *Variety* article reports the efforts of the WB in that year to commission original films intended to target one of the network's key demographics, women aged 18- to 34.[43] Dempsey quotes the WB's senior vice president of original movies, Tara Jamison, as saying, "We want concept-driven movies that the WB can use as stunt programming," and explains the network's goal: to attract audiences who might not otherwise turn to the WB and entice them to stay for the network's regular series programming.[44] In such cases, cable networks seem willing to "forgo immediate profits" of selling time to advertisers in the hopes of using their more prestigious original programming to draw in an audience that might remain beyond the end of the film.[45] Ultimately, it proves difficult to determine whether or not this belief in the "lighthouse effect" has any definitive evidentiary support. Successful made-for-TV movies such as Disney's immensely popular *High School Musical* franchise have demonstrated the ability of the form to draw audiences to a cable channel, but whether those audiences stay to watch the rest of the channel's offerings remains unclear. Nonetheless, enough cable executives seem to believe in the faculty of the made-for-TV movie in this regard to count it among the industrial strategies and attributes of the form.

Undoubtedly, the made-for-TV movie has successfully relocated to cable, where it continues to fulfill important economic functions for channels seeking to develop a particular identity as well as credibility with critics and audiences alike. Just as the form helped to build programming schedules for the broadcast networks, so has it become a staple of cable. But cable networks have not been the only outlets to find industrial utility in the form. Independent producers have similarly discovered the profit potential of increased international co-production of made-for-TV movies, as well as the attendant ease of overseas distribution.

The Made-for-TV Movie on the International Market: Changes to Content and Financing

The rise of independent international distribution companies in the 1970s, the 1980s, and into the 1990s led to the increased viability of selling long-form television—such as the made-for-TV movie—in international markets. McMurria describes the rise of international distributors in the 1970s as related to the introduction of the Financial Interest and Syndication Rules in 1972.[46] Companies such as Saban, World International Network (WIN), and Hamdon

Entertainment developed as the means for American producers to syndicate their programming abroad and earn additional revenue. As these corporations grew in size and influence in the 1980s, television producers found it increasingly financially practicable to develop programming intended to appeal to an international market. By the 1990s, this shift from a particularly domestic focus to producing programming intended for both a domestic and an international audience ultimately resulted in several changes to long-format television programming.

The made-for-TV formula that appeared on broadcast networks in the 1970s and 1980s featured themes tied to the dominant headlines of the time. This created a particular cultural resonance for the American market and was typically targeted toward a female audience, but it was not appreciated by international distributors because of its incongruity with what they perceived the viewers in their markets to desire. McMurria notes the reluctance of these distribution companies to acquire and market such films overseas, citing commentary by Larry Gershman, founder of WIN, who explained that "the genre of original movies most prevalent in the late 1980s, those focused on women protagonists and family relationships, were less marketable internationally."[47] McMurria points to a tendency on the part of these distributors to rename made-for-TV films, ostensibly in order to make them more appealing within a global marketplace, for instance changing *The Price She Paid* (1992; the story of a rape victim) to *Plan of Attack* in order to make the film seem more like an action thriller.[48] Unfortunately, the social relevance that had become a hallmark of the American made-for-TV film did not make for a saleable product overseas, a problem McMurria explains with the comments of Hallmark Entertainment producer Robert Halmi, Sr.: "There is a whole genre of American programming that cannot work overseas. It is too headline-oriented: the sickness of the month, the murder of the month. That is simply not translatable."[49] The problem was not with the topic itself—AIDS, nuclear warfare, and domestic violence are unfortunately universal—but the way in which these stories were told appealed to a particularly American audience. Those purchasing the films for audiences outside the United States considered the way the story unfolded and the characters developed to be specifically American. For example, part of the cultural relevance that made *Brian's Song* so compelling in the United States was viewers' familiarity with the title character as a popular athlete and with the particularity of US race relations. Certainly viewers in other cultural contexts could draw parallels (the death of a local athlete who overcame relevant social divisions), but drawing these parallels required more work on the part of audiences than other available programming options.

Ironically, the relevance and topicality that made the made-for-TV movie successful in its inception contributed to its downfall in the post-network era once increasingly cash-strapped networks sought to diminish license fees and as the content that led to success in the US market narrowed its appeal abroad. Socially relevant content made the programming too American, and the films had to adjust their content in order to continue in the increasingly necessary international markets. In order to produce a television film suitable for international sales, the made-for-TV movie began to look more and more like the theatricals that served as the initial inspiration for the form. Rather than being like the social issue films so popular in the 1970s, more recent made-for-TV films have shied away from melodrama in favor of more action-oriented narratives, as the perception of distributors that programming emphasizing spectacle over dialogue would be more popular and saleable overseas.[50] As Barbara Selznick notes, "The action genre (and its hybrids—western action, historical action, biblical action) became most popular for co-productions."[51]

Shifts in casting and the telling of more universal (or at least less distinctly American) stories augment other content-related changes that have occurred as a result of the increased internationalization of made-for-TV movies. One part of the strategy of making the films more successful overseas involves using non-US actors, who are more likely to appeal to a global audience. Another strategy involves featuring stories that are more universal than particular. Films telling international history, such as *Joan of Arc* (1999) and *Nuremberg* (2000), have become more common, as well as classic tales such as *The Odyssey* (1997), *Jason and the Argonauts* (2000), and *Arabian Nights* (2000). Still, reaching such a broad audience is quite challenging. As the *International Herald Tribune* notes, "The scramble to please all the world's viewers all of the time, or as many and as consistently as possible drives some producers to distraction."[52]

In addition to changing the content, casts, and titles of made-for-TV films in order to appeal to international audiences, the increasing importance of international trade alongside expansions in protectionist regulations led to growth in the international co-production of made-for-TV films. International co-production deals involve alignment among producers and distributors in various countries to share the expenses and profits of production. Television films for the US market have increasingly been produced with foreign companies and stars in order to make the films more likely to appeal to a broader international audience and be capable of surmounting cultural trade barriers.

Selznick cites Carl Hoskins, Stuart McFayden, and Adam Finn's concept of "cultural discount—the loss of value faced by a cultural product when it leaves its country of origin based on the fact that it will not be fully understood or appreciated by the new, culturally distinct audience."[53] Cultural discount, then, must be seen as a distinct possibility for American-produced made-for-TV movies exported overseas. But, as Selznick notes, the impact of this cultural discount can theoretically be lessened if a production has creative input from another culture. As she explains, a co-production "with a French partner may seem more 'French' than if the partner had not been involved."[54] The concept of "cultural discount" has clear economic implications, as distributors in many countries are willing to pay more for programming that seems to relate to their own culture. Pointing to Sharon Strover's example of the American, Canadian, and French co-production *Scene of the Crime*, Selznick explains that the series garnered around $200,000 per episode in France when American series were earning only $40,000.[55] This, she claims, is due to the fact that the cultural discount of the co-production was less than that of an American-produced series. Moreover, the series could be labeled French and thus operate within the restrictions of the French programming quota.

International co-production guarantees an overseas market, as countries that have aided in the production of programming are understandably interested—indeed, have invested—in distribution within their own markets as well. These international co-producers allow American production companies to bypass regulations stipulating that only a certain percentage of programming can be imported. One example of such restrictions is the European Union's 1989 "Television without Frontiers" directive, a regulation requiring that a majority (minimum 51 percent) of any programming shown on television in Europe must be of European origin.[56] Many European countries further strengthened these quotas with requirements of in-country production as well. France has the most stringent limitations; a *Financial Times* article reports, "French TV stations are obliged to carry 60 per cent European-origin programming at prime time, of which two thirds must be French."[57] By co-producing television films with European companies, American producers ensure that their films can be sold and aired in Europe without concern for quotas. A 1997 article from the *International Herald Tribune* explains further causes of the rising presence of international co-production: "What has driven the boomlet in international co-production over the past decade has been the explosion both in overseas television channels and networks and in program production costs."[58] As broadcasting systems outside of the United States continue to open up, their need for programming increases. International co-production serves as a way to meet programming needs while keeping costs down.

International production strategies extend beyond European co-production alliances that have increased production in London, Berlin, and Prague, however. American producers have become increasingly reliant upon moving their production north to Canada, in a practice called runaway production, because production costs are lower and any film produced is considered suitably non-American (at least in part). McMurria notes, "A June 1999 report commissioned by the Directors Guild of America and Screen Actors Guild found that of 308 made-for-TV movies and mini-series developed in the United States in 1998, 139 were economic runaways (compared to thirty in 1990)."[59] Undoubtedly, the lure of tax incentives, cheaper (non-unionized) labor, and the label of international co-production have proven too tempting for American producers to resist.

All of these shifts in both production and content are clearly related to the changes within the industry that occurred within the multi-channel transition. No longer are all films produced by American companies for just an American audience; increased global conglomeration of media has led to an ever more international television marketplace. Once prized for its ability to tell particularly contemporary American stories, the form has since been appropriated as a vehicle through which to address a global audience, a shift that has changed its content, production, and economics.

Conclusion: The Made-for-TV Movie in the Post-Network Era

> TV movies are a vestige of the past … This genre is over.
>
> Jeff Zucker

A 2001 article in the *St. Petersburg Times* heralds "The death of the original network TV movie-of-the-week" and offers the above quotation from then NBC entertainment president, Jeff Zucker.[60] Despite such pronouncements, the form was not then, nor has it ever been, dead and buried. Even the aforementioned article reports that, in the previous year, the same networks that were "over" the made-for-TV movie aired 146 such films. Although this figure represented a decline of 44 percent, it in no way signaled the absence of the form from television screens. Indeed, as vice president of research for Lifetime, Tim Brooks, notes, in an effort to explain what some have argued is a fading of popularity of the form, "I don't think it's that [viewers] are tired of the network's [*sic*] movies. It's that there's so much competition for their attention."[61]

These two quotations, one from a broadcast network executive and the other from a cable channel executive, highlight the two sides of the made-for-TV movie in the post-network era. Despite the fact that the form helped networks build an audience during the 1970s and 1980s, the changes caused by the arrival of cable and the realities of the multi-channel transition forever altered the economics of the television industry. No longer were these films able to draw a large enough audience to justify their presence on broadcast network schedules. Nonetheless, although executives, audiences, scholars, and critics alike continue to declare the form dead, in truth it enjoys a successful and abundant afterlife on cable and abroad. The industrial features of the made-for-TV movie that made it so successful—its relative cheapness and its status as fleeting special event programming—are the same aspects that have allowed it to survive the changes to the industry throughout the multi-channel transition and continue now at the beginning of the post-network era.

Despite the adjustments to the made-for-TV movies that developed during the multi-channel transition, the key features of the post-network era do not seem to have a great deal of affect on the form. In fact, even the few changes that have any impact at all only serve to strengthen their position within the industry. Technological changes within the post-network era, including platform changes (the ability to watch television on your iPod or mobile phone), digital cable, and the digital video recorder, have not had tremendous consequences for the made-for-TV movie, either textually or industrially, although they have allowed the "dead" form to enjoy even greater viewership, as interested parties can record or download the films to watch at their convenience.

In terms of changes to creativity and production in the post-network era, the opportunity for greater aftermarket value has also worked in the movie's favor, allowing for the possibility of sales beyond even second- and third-run network airings. Indeed, as nostalgia for the most memorable made-for-TV movies of the past increases, so too does their financial viability in aftermarkets. Lifetime has attempted to capitalize not only on nostalgia, but also on the desires of audiences to participate in their own creative work through their online "Lifetime Movie Mash-up" program.[62] This offers participants a large array of scenes from Lifetime made-for-TV movies that can be edited and compiled to create a new "mash-up." The scenes available are some of the network's most melodramatic and salacious, and some of the "Top Mashed" listed bear titles such as "Sex!" "Sex Sex & Sex" and "Baby Drama"—reflections of the form's sensational roots. The Lifetime Movie Mash-up provides an excellent example of a cable channel capitalizing not only on its melodramatic original programming but also the post-network era element of increased amateur production.

Although it is considered a "vestige of the past," there can be no denying the made-for-TV movie's importance within the television industry. The development of the form, far from dead, was clearly affected by the multi-channel transition. While the creation of cable may have led to conditions that diminished its place on broadcast networks, the movies proved well suited to cable channels' narrowcast environment and need for programming that provided branding and prestige. Shifts in the global television marketplace and the increased reliance on international co-production similarly have helped to define the made-for-TV movie within the multi-channel transition and now the post-network era.

Notes

1 Todd Gitlin, *Inside Prime Time* (New York: Pantheon Books, 1983), 157.

2 As they disappeared from broadcast airwaves, they also became increasingly invisible in media scholarship; to draw attention to this I offer the dates of the scholarship I cite in text to historicize it properly within the existing scholarship on the made-for-TV movie, which has been negligible in the last fifteen years.

3 NBC's *See How They Run*, which aired on the network October 7, 1964, is generally considered to be the first made-for-TV movie.

4 Elayne Rapping, *The Movie of the Week* (Minneapolis: University of Minnesota Press, 1992), 12.
5 Gary Edgerton, "High Concept, Small Screen," *Journal of Popular Film and Television* (Fall 1991).
6 Ibid.
7 "Telefeature" here refers to the term for films developed for television but more closely related to Hollywood feature films. "Docudrama" refers to the programming most closely aligned with the melodramatic made-for-TV movie we often consider exemplary of the form today.
8 Douglas Gomery, "Movies on Television," *Encyclopedia of Television*, http://www.museum.tv/archives/etv/M/htmlM/moviesontel/moviesontel.htm.
9 Rapping, *The Movie of the Week*.
10 Ibid., 19.
11 Laurie Schulze, "The Made-for-TV Movie: Industrial Practice, Cultural Form, Popular Reception," *Hollywood in the Age of Television*, ed. Tino Balio (Boston: Unwin Hyman, 1990), 365.
12 John Dempsey, "Cablers Pony up for Pics." *Variety*, 7–13 August 2006.
13 Elizabeth Jensen, "Miniseries Find New Niche After Nets Flee; Cable Longforms Grab Emmy Nods and Audiences," *Television Week*, 31 July 2006.
14 James P. Forkan, "Networks Trim Movie Slates," *Advertising Age*, 27 July 1987, 53.
15 Ibid.
16 Stephen Farber, "'88 TV Season: Mini-Series and Films," *New York Times*, 18 August 1988, 22.
17 Ibid.
18 Monica Collins, "NBC Films: Skimpy as a Bikini," *USA Today*, 17 February 1989, 3D.
19 Ibid.
20 Rick Kissell, "True Tales Dominate Telepix Turf," *Daily Variety*, 14 September 2001.
21 John Dempsey, "Cable's Mad for Made-Fors," *Variety*, 26 July 1999, 17.
22 "Hallmark Channel," *Advertising Age*, 14 April 2008.
23 Dempsey, "Cable's Mad for Made-Fors," 17. Emphasis added.
24 Paul J. Gough, "'Knight Rider' Revs NBC's Ratings," *Associated Press*, 19 February 2008.
25 Ibid.
26 Schulze, "The Made-for-TV Movie," 364–5.
27 John Dempsey, "Cablers Split on Made-For Strategy," *Variety*, 11–17 October 2004.
28 Jon Lafayette, "Originals Lift Cable's Profits, Reputation; Nets See Return on Record Investment in Summer Lineup," *Television Week*, 17 September 2007.
29 Paula Bernstein, "Basic Cable Stakes Telepic Terrain," *Daily Variety*, 14 June 2004.
30 "Ratings – Sept. 29–Oct. 5," *Broadcasting & Cable*, 13 October 2003.
31 Ibid.
32 Dempsey, "Cablers Split on Made-For Strategy."
33 John McMurria, "Long-Format TV: Globalisation and Network Branding in a Multi-Channel Era," *Quality Popular Television*, ed. Mark Jancovich and James Lyons (London: British Film Institute, 2003), 70.
34 Ibid., 72.
35 Eileen Meehan and Jackie Byars, "Telefeminism: How Lifetime Got its Groove," *Television and New Media*, 1, 1 (2000), 33–51.
36 Classically melodramatic offerings from Lifetime include *A Face to Kill For* (1999), *A Face to Die For* (1996), *A Job to Kill For* (2006), *A Vow to Kill* (1995), and *A Friendship to Die For* (2000).
37 Amanda D. Lotz, *Redesigning Women: Television After the Network Era* (Urbana-Champaign: University of Illinois Press, 2006), 53.
38 Ibid., 55.
39 McMurria, "Long-Format TV," 72.

40 Ibid.
41 Jensen, "Miniseries Find New Niche."
42 Bernstein, "Basic Cable Stakes Telepic Terrain."
43 John Dempsey, "Frog Gets in Movie Mood," *Variety*, 28 June–11 July 2004.
44 Ibid.
45 Jensen, "Miniseries Find New Niche," 74.
46 McMurria, "Long-Format TV," 74.
47 Ibid., 76.
48 Ibid.
49 Ibid., 66.
50 See Timothy Havens, "'It's Still a White World Out There': The Interplay of Culture and Economics in International Television Trade," *Critical Studies in Media Communication*, 19 (2002), 377–98, for an examination of the complex dynamics of how the perceptions of international buyers and sellers affect the creation of television.
51 Barbara Selznick, *Global Televison: Co-Producing Culture* (Philadelphia: Temple University Press, 2008), 11.
52 Ibid.
53 Ibid., 19.
54 Ibid.
55 Ibid.
56 "Television Broadcasting Activities: 'Television without Frontiers,'" *Europa: Activities of the European Union: Summaries of Legislation*, http://europa.eu/scadplus/leg/en/lvb/l24101.htm, accessed 1 December 2007.
57 David Buchan and Emma Tucker, "France Steps up to TV Pressure," *Financial Times*, 3 February 1995, 2.
58 Richard Covington, "Tailoring Film and TV for the World," *International Herald Tribute*, 2 October 1997, 11.
59 McMurria, "Long-Format TV," 78.
60 Eric Deggans, "Made-for-TV Makeover," *St. Petersburg Times*, 27 May 2001, 8F.
61 Ibid.
62 The "Lifetime Movie Mash-up" program can be found at http://moviemashup.mylifetime.com/.

Chapter 8

The Dynamics of Local Television

Jonathan Nichols-Pethick

The history of broadcast television in the United States has been written primarily as the history of network television and network-derived programming. Even as we transition into a post-network era, the major networks have continued to be the centerpieces of critical and public attention.[1] The Big Three networks (ABC, CBS, and NBC), along with FOX, then UPN and the WB (later merged and renamed the CW), get the lion's share of attention in light of their high profile, prime-time programming and (in the case of the Big Three) large, historically and culturally significant news divisions. The centrality of networks in the critical and public imagination should come as no surprise. Despite recent inroads made by cable networks, programs owned, financed, and distributed by the major networks still claim higher ratings than programming on cable networks. As recently as January 14, 2009, the top-rated cable network program was an episode of *Monk* on the USA Network, which ranked 55 out of 100, with a 3.1 rating for total households. The top 54 shows, and all but seven of 56 through 100, aired on a broadcast network.[2] What this means is that more viewers in the US experience television via the major broadcast networks than by any other means. What is too often lost in this equation, however, is that most of this network television is accessed via local stations affiliated with the networks in question.

Local stations should be integral to an examination of broadcast television in the post-network era for several reasons. Most obviously, these stations hold the local licenses granted by the FCC and are, thus, on the front lines of all debates regarding the public interest. Additionally, the emergence of the new technologies and delivery platforms that currently define the contemporary media landscape introduce a range of opportunities and threats for local stations. For example, the use of broadband technologies and the Internet open up opportunities to restructure how they approach the very idea of localism by becoming key portals of a wider range of community information. At the same time these technologies threaten the centrality of affiliates in the national broadcasting equation by creating avenues for networks to develop new distribution systems that do not require locally situated stations to deliver their signal to customers. Finally, stemming from the previous example, local stations' relationships to networks, advertisers, syndicators, and regulators will likely change dramatically over the next decade. Attending to the economic, technological, and regulatory developments that affect local stations are crucial components of fully understanding the post-network period.

This chapter seeks to examine the range of new and emerging forces affecting the fabric of local television. Several challenges confront any project about local television, however. First, the sheer number of local stations in operation across 210 local "Designated Market Areas" (DMAs) makes it virtually impossible to account for all of the individual variations that might define the local television landscape: in 2007, 1,364 stations operated in the United States (780 of these were UHF stations, while 584 were VHF).[3] Second, there are significant variations in market size. Typically, local DMAs are organized into five primary market categories: the largest, major metropolitan markets (1–25), the mid-range markets (26–50; 51–100; 101–150), and the smallest markets (151–210).[4] In terms of budgets and revenue streams alone, large and small market stations function in very different ways, making any easy generalizations across them difficult. As an example of these disparities, the Project for Excellence in Journalism reports that stations in the top 25 markets average $80 million in annual revenue while those in the next bracket (26–50) average just under $30 million and those in the smallest markets (151+) are just under $5 million.[5] Finally, another way in which stations differ from one another is in their relationship to the large broadcast networks. The first and most obvious distinction is between those stations that are affiliated with broadcast networks ("affiliates") and those that are not ("independents"). This distinction has mattered less in the last two decades, since the emergence of FOX, CW, and PAX as additional networks in need of affiliates has reduced the number of independents. The role of independent stations may come to matter more in the future, however, as the very necessity of affiliation comes up for question. Given the current dominance of affiliates as the norm of local station operation, though, the chapter focuses on these rather than on independents.

Still, key differences exist just within the group of stations affiliated with networks. A majority of the stations in the largest markets are owned by the major broadcast networks (these are referred to as "O&Os," or owned-and-operated stations). These form the backbone of network operations and account for a disproportionately large and desirable share of the viewing audience: that in large urban areas and their surrounding suburban communities. Stations in medium and small markets are typically owned by station groups (such as LIN, Sinclair, Nexstar, or Belo) and generate the vast majority of their revenue from advertising sales during local news. These stations operate at a relative disadvantage to network-owned stations, especially in a political environment sympathetic to deregulation of ownership caps. For example, since local stations rely heavily on syndicated material to fill out their programming schedules and lead viewers into their local news, more ownership power in the hands of the large media conglomerates that produce much of the most popular of this programming, such as Disney, NBC Universal, or News Corp., could create an unbalanced playing field for access to these valuable programs (see this volume, chapter 3). In 1998, for instance, all 22 stations owned and operated by FOX were awarded the syndication rights to the 20th Century Fox-produced series *King of the Hill* before bidding on the series was opened up in every DMA. By pre-emptively awarding the show to its corporate partner stations, 20th Century Fox was able to reach 40 percent of the country and the majority of the major markets with the series, virtually guaranteeing its success and raising its price tag.[6] These intra-conglomerate deals raise concerns about unfair syndication practices favoring producers, networks, and stations that are part of the same corporate family, and underscore the categorical differences between stations owned by the networks and those that are not.

Despite their differences, economic fluctuations, technological developments, and regulatory modifications affect all local stations in one way or another. Like the major networks, local stations are struggling to reinvent themselves within the context of rapid technological, regulatory, and structural ferment. But they do so in an environment in which, unlike the networks, their very existence is at stake. In order to survive the range of changes that define the post-network era, local stations need to think of themselves in a truly post-network sense: as having to find ways to exist as independent operators serving local communities in a manner that has very little to do with being the local franchise of a national network chain (at least as that has been defined in the past, i.e., via a contractual relationship with a major broadcast network). Ownership trends of large station groups, renewed regulatory interest in issues of localism, changing financial relationships with networks, general declines in overall audience share, and the promise of new revenue streams via emerging digital technologies all come together to make the fate of local television stations quite uncertain.

Affiliates in the Post-Network Era

Understanding the state of local television in the post-network era requires that we address changing conditions in three primary areas that dictate the daily functions of local stations: industry economics, technology, and government regulations. Of course, these areas are deeply interconnected—each one affecting the others in several ways—but there is heuristic value in breaking them apart for analysis. In each case, analysis reveals contradictory trends: a complicated dynamic that can be most simply understood as the interplay between expansion and contraction. Eric Klinenberg notes this particular dynamic with regard to the concerns over ownership regulations:

> These opposing tendencies help explain one of the great paradoxes in today's media business: while citizens are bombarded with an apparently endless supply of media products, the shrinking supply of primary producers at the local level renders media content strikingly similar, regardless of its form.[7]

The post-network media industries in general are defined by this particular dynamic: as large conglomerates continue to *expand* their holdings and invest in new delivery technologies, their focus on the audience has *contracted*, aiming at narrower and more specific "quality" demographics, and their news and production operations have been restructured and tightened in order to emphasize cost-efficiency.

This somewhat contradictory dynamic is central to the realities of a changing electronic media landscape in which large broadcast networks and the local affiliates with which they partnered once enjoyed a near oligopoly in terms of programming and access to audiences. While large diversified media conglomerates still enjoy a similar control over content and access to audiences, individual outlets, such as television networks and local stations, can count on only a fraction of the audience they once commanded. For example, ratings for the top twenty programs in 1972 ranged from 34 (*All in the Family*) to 21.9 (*Bonanza*).[8] In contrast to those numbers, for the week of March 24–30, 2008, the top-rated show (*American Idol*) recorded a rating of 17.2 while the tenth highest rated broadcast program (*Extreme Makeover: Home Edition*) received a rating of 8.5. Maximizing revenue under these conditions requires flexible strategies for capturing particular "quality" audiences across multiple outlets.

As Michael Curtin has pointed out, in the post-network era (or *neo*-network era, to use his term) these conglomerates actually deal in two kinds of media products simultaneously: more "traditional" mass-oriented forms that have expansive (i.e., national, perhaps even global) appeal, requiring less intense investment (as in the case of covering the Olympic Games), and those that are targeted at small niche audiences with a great deal of personal investment in the product, as evident in cable programs such as *The Daily Show* on Comedy Central.[9]

The dynamics of industrial expansion and contraction and tensions between mass and niche targeting that partially define the post-network era are not just the province of large conglomerates; they affect local media outlets as well. In the rest of this section, I explain how these dynamics play out across economic, regulatory, and technological developments of local television stations.

Economic Trends for Local Television

Several new trends inform the current status and functioning of local television stations in general. Of particular importance are the changing ways in which stations are trying to generate and expand revenue streams amid threats of a shrinking audience base. Three central factors in understanding affiliate station economics are their relationship to their network partner, their relationship to local cable and satellite franchises, and their ability to continue expanding local advertising sales across new platforms such as the Internet and new digital broadcast spectrum.

The primary economic relationship between affiliated stations and partner networks—the system known as "compensation"—has changed considerably in the past decade, essentially disappearing. During the classical "network era" of television, networks compensated affiliated stations for access to their locally licensed spectrum. Networks were paying ostensibly for access to local audiences during certain dayparts (morning news, mid-day soap operas, prime time, and late night) that they could then, in turn, "sell" to national advertisers.[10] For local stations, compensation has been, until recently, an important part of their overall revenue structure. In 1990, CBS paid its 200 affiliated stations a total of $121 million in compensation, down from $150 million the year before.[11] By 1999 NBC was paying almost $200 million to its affiliate stations.[12]

The system of compensation began to unravel in the mid-1980s, when each of the networks was purchased by a larger company: NBC by GE, ABC by CapCities, and CBS by Westinghouse. The financial management strategies of these larger corporations were not necessarily tied to or respectful of traditional relationships within the television industry. With cable and satellite penetration growing rapidly, these new corporate owners saw little reason to hold on to economic relationships that had been established under a different set of historical and institutional contexts in order to compete in a multi-channel environment. In 1986, Lawrence Tisch threatened to drop affiliations if the stations didn't agree to reduce compensation agreements in light of CBS's financial struggles; then, in 1992, CBS announced that it was going to alter compensation agreements with affiliate stations significantly, even to the point of charging the stations to carry certain programs.[13] These changes were announced as a necessary adjustment to the "fundamental procedures of the broadcasting business to respond to the new competition" from cable and satellite, FOX's emergence as a powerful fourth network, and syndicators' willingness to pay for carriage on local stations.[14] At the time, Anthony Malara, president of affiliate relations for CBS, called the move "the most dramatic change in the economic relationship between a network and its affiliated stations in 25 years."[15]

Since that time, the networks have not only reduced compensation to stations, they have mostly eliminated compensation altogether. In 2000, Hearst–Argyle signed a ten- year deal for its ten NBC affiliates that "gradually takes its compensation down to zero."[16] NBC announced in July of 2008 that it will ask stations to pay reverse compensation when their contracts with the network come up for renewal, meaning that stations will now pay the network for access to its programs.[17] And FOX initiated reverse compensation in the late 1990s, requiring stations to pay the network for access to programming when ratings exceed expectations.[18] These initiatives from NBC and FOX may appear to expand CBS's early 1990s policies, but they more accurately indicate a radical revision of the economic relationship between networks and their affiliates.

The networks cite a range of factors in their decision to cut compensation and even request payments from affiliates—in particular, increasing production costs of prime-time series and the rising price of exclusive rights to major sporting events such as the Super Bowl, World Series, and Olympics. Since the mid-1980s takeovers, the networks have positioned themselves as besieged by hard economic times, increased competition, and underperforming local stations. According to the networks, competition has expanded, audiences have fragmented across a broader range of available programming and platforms, and costs for major sporting events and hit shows have skyrocketed. At the same time, however, the long pattern of compensation reductions points to the continued, even growing, power and leverage of the networks. Stations are eager to maintain their status as affiliates with the threat of cable and satellite bypass on the horizon and the value of stations still tied to their network brand.[19] The price for losing a network affiliation can be steep, as in the case of KRON in San Francisco, which lost its NBC affiliation in 2002 and saw a 40 percent decline in its ratings.[20] Thus, while the term "post-network era" might encourage discourses about decreased power and control for the networks, it more accurately signals a shift in the primary relationships that have defined network television for decades—a process that began in the mid-1980s and has resulted in a near complete reversal of the economic terms on which the networks and their affiliates relate to one another.

A second economic factor affecting local stations also involves a reversal of sorts in their economic relationship with the cable systems that deliver their signals to more and more homes. Whereas the duels over station compensation by the networks point in the direction of decreased leverage for stations, the opposite is true of their dealings with cable providers over the issue of retransmission consent fees. These fees refer to money paid by individual cable providers to content providers such as cable networks and local broadcasters. They are paid per-subscriber, per month, and range from nearly $3.00 at the high end (ESPN) to just pennies at the low end (the National Geographic channel). In coarse economic terms, the greater the demand for a particular channel, the higher the fee. Demand for local channels is necessarily high, as they are the most direct source for breaking local news and weather information. Until recently, however, cable has been the only way, other than using a television antenna, to access local stations and has therefore enjoyed significant leverage in any negotiations with stations. But with the growth of satellite carriers offering local carriage, as well as competitors such as Verizon and AT&T entering the market, cable operators now find themselves with significantly less leverage.

Although cable networks such as ESPN, TNT, and MTV have always received these fees from cable providers, local broadcast stations have historically been offered a "must-carry" option that requires the provider to carry their signal but does not require a payment from the provider to the local station. "Must-carry" initially resulted from an FCC mandate in 1972 that helped protect local broadcasters as cable began to expand from its earliest manifestation as Community Antenna Television, which was simply a way of getting local broadcast signals to hard-to-reach areas of the country.[21] With an increasing number of cable networks competing for limited bandwidth, the must-carry rules were designed to insure that local signals were included in any cable package.[22] In 1994, with competition from satellite delivery services growing, Congress authorized stations to open up "cash for carriage" negotiations with cable providers. From the stations' perspective, cable operators are already paying cable networks such as ESPN and Lifetime for the rights to distribute their content and should treat local stations in the same manner. Cable operators argue that broadcast signals are already free to consumers (since most of them can receive these signals with an antenna) and so shouldn't be monetized in the same way as cable-only signals that can't be picked up over the air at the consumer end. Stations see this redistribution of their content without payment as a clear violation of copyright. Cable operators were not particularly threatened by the prospect of payment until satellite services could work out the technical and contractual details of carrying local signals, since they were still one of only two primary means of receiving these valuable signals.

The value of local signals, ironically, has turned against cable operators. In 1999, Congress passed the Satellite Home Viewer Improvement Act (SHVIA), which allowed satellite operators to offer local signals to their customers. Before this legislation, satellite providers had been barred from including local signals as part of their service. The rationale for this ban was to protect local stations from the potential of satellite providers importing redundant local signals from distant locations. Lifting the ban in 1999 was a matter of putting satellite on equal competitive footing with cable, though, under the current law, satellite providers are required to pay for the local signals they offer. Once satellite included local stations and offered a truly comparable service, satellite penetration increased and has steadily diminished cable's subscriber base.[23] Broadcasters now argue that the norm of paying for carriage common in satellite distribution should also serve as the norm in cable distribution.

An added pressure on cable comes from expanded broadband capabilities and looser government regulations on cross-ownership. Together, these developments have encouraged other telecommunications companies, such as Verizon and AT&T, to invest in local video delivery services, and they too pay local broadcasters for the right to distribute their signals. Having these alternative delivery systems as potential service options for viewers adds leverage to local stations in their negotiations with cable providers. And, as cable providers also move to offer broadband Internet access and phone services, the potential loss of customers to alternative television delivery systems threatens their broadband and telephone businesses as well. All of these developments taken together have changed the dynamics of the playing field to favor local stations in their negotiations with cable providers.

The reasons broadcasters are now so willing to demand retransmission fees stem from several factors. First, as already discussed, the increased local competition from satellite and telecom services that are already paying broadcasters these fees diminish the leverage of cable systems as the only other service option. Second, those stations that are not owned and operated by a broadcast network are under increased pressure to replace the lost compensation revenue. Third, and related to the second point above, stations have been forced by the FCC to invest heavily in digital and HD technologies, and they need to find ways to recoup that money. Finally, stations are now more likely to be owned by a larger station group with greater negotiating leverage. These groups can pull multiple stations from a single cable provider across a range of markets, creating a broader reach of customer dissatisfaction if the cable station loses the ability to retransmit the local station.

In January 2005, one of these station groups, Nexstar, pressed the retransmission consent issue when its contracts expired with two multiple systems operators (MSOs): Cable One and Cox Communications. After the two MSOs refused Nexstar's demand for $0.30 per subscriber, Nexstar pulled its signals from these systems for nearly a year—in other words, subscribers in places such as Joplin, Missouri, Shreveport, Louisiana, and Abeline, Texas, found themselves without cable access to at least one network affiliate station. To get a clearer sense of the financial stakes involved, Cox estimated that payment to Nexstar in just a handful of markets would amount to $1.3 million annually.[24] Nexstar eventually won the battle as the two sides reached an agreement for fees in December 2005, establishing a precedent that has changed the terrain of the broadcast–cable relationship. In January 2007, Sinclair Broadcasting, a larger group than Nexstar, followed suit and pulled 22 of its stations from Mediacom cable systems, Belo pulled stations from Charter systems in five markets, and Northwest Broadcasting pulled its FOX affiliate in Spokane off the Time Warner system there. Currently, the contracts of the third largest station group—CBS Corporation—are coming up for renegotiation. And Les Moonves, president and CEO of CBS Corporation, has publicly stated his intention to seek retransmission fees in these negotiations. Given the current trend established by smaller players such as Nexstar and Sinclair, the windfall for CBS could be significant in that it will test the limits of what the largest MSOs such as Comcast are willing to pay.

Finally, a third economic indicator for local stations in the post-network era involves the strength of local advertising in terms of both traditional television spots and the growth of online platforms. Local television has been the top television advertising market in the United States since 1964, according to the Television Bureau of Advertising. In 2007, local stations accounted for $24.5 billion in advertising sales. Of that total, $14.4 billion came from local spots, while national spots totaled $10.1 billion. The total advertising dollars for local television stations accounted for 35 percent of all television advertising sales nationally, which was down slightly from 36 percent in 2005.[25] This slight decline in advertising revenue might be explained in part by the fact that ratings for local newscasts, which are the primary generators of such revenue for local stations, were down anywhere from 3.3 percent to 7.4 percent in evening and late news slots, with only morning newscasts holding steady or making slight gains.[26] Thus, while local television remains the most lucrative advertising market in television, stations are seeing signs of decline in traditional viewership that will likely continue as more viewers begin to migrate to online distribution platforms for their news and entertainment.

In respect to the decline in traditional viewing patterns, local stations have made significant investments in their online presence. These investments are in the form of expansions in sales and technical staff, though very few stations have invested in additional staff to produce web-only content. The trend as of 2008 is simply to repurpose traditional content for distribution via the stations' websites or alternative distribution sites such as YouTube. Advertisers are following stations and viewers online as well. The Television Bureau of Advertising estimates that online ad sales at local stations will reach $978 million in 2008, which represents an increase of 42 percent from the previous year. Clearly, the most significant economic growth in local television now and for the foreseeable future lies in the development of web-based distribution channels and advertising platforms.

As the previous examples show, the economic trends for local television in the post-network era are fraught with contradictions. Local stations find themselves in positions of weakness with regard to their partner networks and strength with regard to MSOs. Their traditional television ratings and advertising sales are shrinking while their online presence and revenues are growing rapidly. Local television signals are valuable additions to any MSO looking to add subscribers to its base, but too narrow a focus on those signals alone could, according to some industry analysts, spell trouble for the stations themselves. As with the larger culture industries in general, the economic future for local stations will require flexibility and attention to multiple, sometimes contradictory strategies for attracting viewers/users and generating revenue.

Technological Trends for Local Television

As indicated above, the future of local television in the post-network era hinges in part on the way stations adapt to and take advantage of emerging digital distribution platforms. These new platforms, however, represent both opportunities and threats. While the Internet represents an area of significant growth for local stations, as discussed in the previous section, it also represents a potential threat in terms of the future of their relationships to partner networks, which makes the future of a television system in which network programming is delivered free of charge to viewers via affiliated local stations increasingly less probable. The broadcast networks that once relied on interconnected local stations to distribute their signals nationally no longer need that technology to access a national audience. The combined availability of cable and satellite systems along with viewing portals via Internet sites could allow networks to bypass the stations altogether should they desire. As Michael Eisner, former chairman and CEO of Disney, told a gathering in 1998:

> ABC eventually will have to change its relationship with its affiliates and eventually it will repurpose programming and eventually our soap operas will be part of a separate soap opera channel and eventually news will be repurposed and eventually we'll multiplex as the technologies and our contracts [with affiliates] come up.[27]

Threats of cable, satellite, and broadband bypass stem not only from technological developments, but also from disputes over key changes in network–affiliate contracts, especially with regard to the compensation and clearance issues. Networks could find themselves increasingly interested in forgoing broadcast affiliation as stations push back at networks' demands for decreased or reverse compensation agreements and allowance of direct competition from repurposing. This possibility is especially salient in light of the fact that combined satellite and cable penetration is over 80 percent.

With regard to cable bypass in particular, there is an irony in the fact that the early development of cable technology was seen as a boon to networks and a potential threat to local stations. Community Antenna Television was capable of delivering network signals to remote and hard-to-reach areas of the country, but it was also capable of delivering distant signals into locations that already received local broadcast service. Early FCC regulations of cable specifically protected against these kinds of market incursions as a way to safeguard the interests of local broadcasters. At that time, the networks were still relatively young and, of course, *television* itself was but an infant; networks needed healthy local stations in order for television to continue to grow. More than 60 years later, however, the incentive for growth in terms of building a national infrastructure has been replaced by new incentives driven more by the promise of personalized technologies than the need to maintain a commercial broadcasting system with national reach: a matter of depth of reach as opposed to breadth.

The threat of cable, satellite, and broadband bypass, while ever-present, remains only a possibility. Networks are still too heavily invested in station ownership themselves to make such a switch, and these stations have spent billions of dollars preparing for the changeover to digital broadcasting in 2009. One of the most obvious opportunities of the digital spectrum is its capacity for multicasting—dividing the allocated spectrum to operate multiple (up to four) broadcast channels simultaneously. In addition to their regular broadcast schedule, some stations now offer a 24-hour news and/or weather channel, a local sports outlet, or additional syndicated programming. The potential of multicasting, of course, is for increased local advertising revenue and possibly multiple retransmission fees. More tellingly, there are a growing number of new networks designed specifically to take advantage of the possibility that stations may want to multicast but cannot invest the time or capital to increase production. Networks such as Dot Two, Ion Life, Qubo, and Trinity Broadcast Network offer 24-hour programming to stations, acting in much the same way as program syndicators, but selling an entire tier of unified programming. Given the opportunities for revenue sharing that these multicast networks offer, it is reasonable to expect that this would be an area of growth for local stations.

The other great technological possibility for stations lies in the potential for Internet and mobile platforms to provide on-demand services for viewers/users. As with other emerging strategies for using new technologies, a transitional logic permeates both the networks and the affiliates. Affiliates have encountered competition for network programming from the networks themselves, which are increasingly repurposing their programs through their own websites and through direct payment platforms such as iTunes. ABC affiliates criticized the network for leaving them out of negotiations when ABC entered into a deal with Apple Computers in 2005 to make Disney-owned programs available on iTunes only one day after their original broadcast.[28] As a way of addressing affiliate concerns, the major networks entered into deals with their affiliate groups that also allowed local stations to offer network programming on their websites, sometimes streaming the shows live on the websites and then offering them as on-demand programming at the beginning of the following day.[29]

Though these agreements point to at least some kind of continued future for network–affiliate partnerships, these agreements are likely to erode as the networks become more comfortable leaving affiliates out of the equation. Thus, affiliates also need to consider ways to build their brands around their own valuable offerings: local news, sports, and weather. As Duane Lammers, former General Manger at WTWO and former chief operating officer at Nexstar, stated: "It's why we focus so much here on local programming: doing Indiana State, doing high school football games. We do more local news than anyone in the market. You'd better have defined yourself when that day comes."[30] These brands will increasingly involve the development of Internet portals for the community. At least part of the impetus for increasing use of these on-demand technologies is to address declining viewership for traditional local newscasts.

According to the Project for Excellence in Journalism, local news saw a decline in audience for the second straight year in 2007, with only the morning news segments experiencing a small increase or merely holding steady. According to Bob Papper, 97.6 percent of all stations that run local news have a website, and these websites are increasingly profitable.[31] Research supported by the Radio and Television News Directors Association shows that stations that reported losses for their online operations dropped by 1.6 percent while those reporting a profit rose by nearly 9 percent.[32] While they vary greatly according to the size and staff of the station in question, a significant trend among station websites is toward on-demand availability of news video and recorded newscasts. More stations are also relying on live cameras and newscasts for breaking news. While the potential to expand news coverage remains a central possibility for these stations, they have also begun to market themselves as distinct from the networks on these sites. WTWO in Terre Haute, Indiana, is a prime example of such a marketing strategy. Rather than playing up the station's relationship to NBC, WTWO's site is branded as "My Wabash Valley" and strives to act as a portal for a broad range of community information. The strategy at WTWO is part of a larger strategy put in place by its parent company, Nexstar: creating websites about their local community that do not carry the brand of the television station at all.[33] The stations that survive the post-network era, it seems, are those that create a strong brand for themselves that doesn't rely on trying to capitalize on old relationships.

In addition to using websites to attract additional revenue, stations are also looking toward mobile technologies for new distribution and advertising opportunities. Stations will likely find a balance between multicasting and mobile broadcasting that involves using only some of their available bandwidth for multicasting and reserving the remaining portion for mobile in a manner offering more potential for generating revenue.[34] In April of 2008, a group of station owners set up the Mobile Video Coalition through the National Association of Broadcasters (NAB) to look into the possibilities of bypassing cell phone companies and transmitting directly to users' devices. The goal in this venture is to locate a new advertising market that early estimates suggest could be worth as much as $2 billion.[35] Along with the Internet and multicasting, mobile represents an alternative platform for delivering programming and generating advertising revenue at a time when advertising sales at local stations are down: 17 percent in the fourth quarter of 2007 as opposed to the gains seen by syndicators and cable networks.[36]

Clearly, new technologies present both possibilities and challenges for local stations in the post-network era. On the one hand, these very technologies give broadcast networks the means to bypass local stations altogether, creating a severe contraction in each local market as stations are forced to compete with

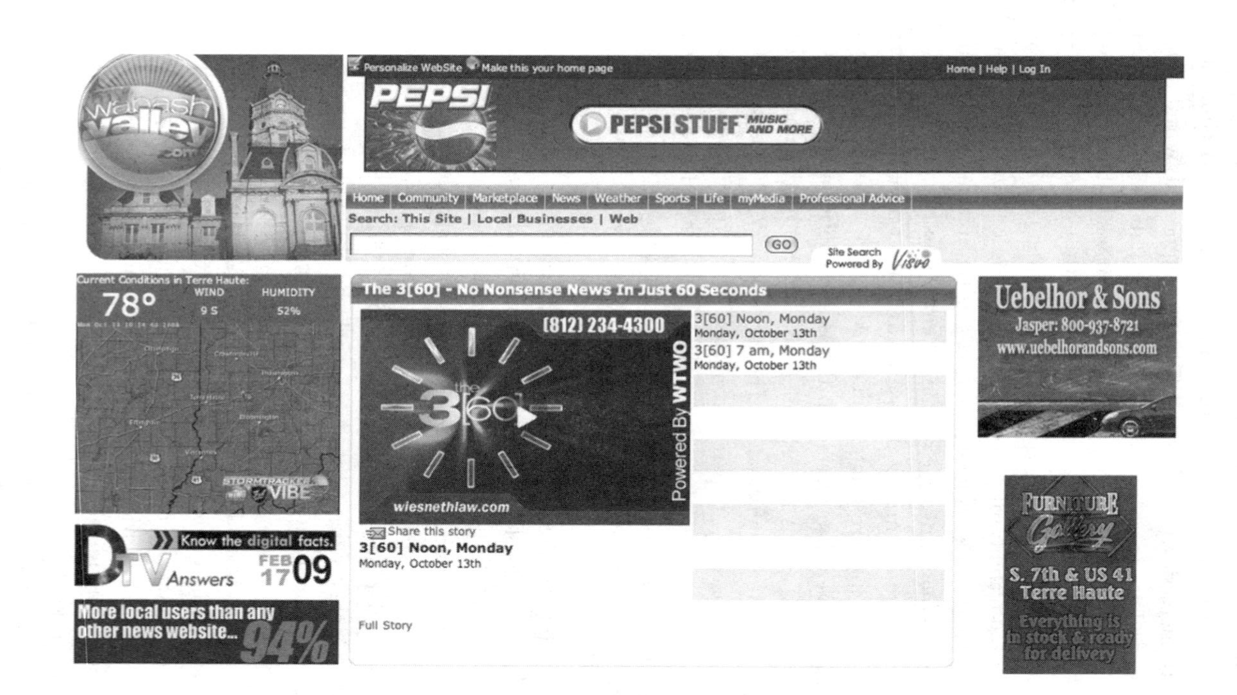

Figure 8.1 This image of the WTWO website (Terre Haute, Indiana) illustrates the many ways the local station attempts to connect its identity with local goods and services. Notice how there is no indication of the station's network affiliation.

Source: http://www.mywabashvalley.com (accessed October 12, 2008).

Figure 8.2 This is one of many local sites with the station logo imposed upon it that the station uses to emphasize its connection with the local community.

Source: http://www.mywabashvalley.com (accessed October 12, 2008).

fewer resources. However, these technologies also allow local stations to accentuate the true value that they do still offer: localism. Competition between stations in individual markets is likely to increase as they work to position themselves as vital portals for local information. These stations may no longer act simply as television stations. Instead, following the trend that defines the larger telecom industry, they will more likely become multi-faceted and central sites for gathering and distributing a wide range of community information: local news, local weather, local sports, and also business directories, real estate guides, and classified ads, to name but a few.

Regulatory Trends for Local Television

Television stations have been regulated within essentially the same structure for nearly 60 years. Regulations restricting ownership of stations have certainly been altered, but the basic structuring tension between free market economics and the government protection of diversity and localism through specific regulations continues. Regulations regarding appropriate content have remained virtually unchanged in spirit, though the specifics of what counts as indecent and/or obscene have necessarily changed along with social mores. Finally, regulations designed to guard localism have been consistent though contradictory in nature: always promoting localism, but never fully enforced.[37] The post-network era, however, presents new challenges for regulators in the face of the kinds of emerging technologies discussed in the previous section. Both content and ownership regulations in the United States are based on the historically and technologically specific condition of "scarcity of spectrum," which insists that a limited resource such as the electromagnetic spectrum, available to all and belonging to the populace, must be protected in the name of the "public interest" from special interests and "harmful" speech.[38] However, significant increases of bandwidth via the digital spectrum and the emergence of on-demand Internet and mobile distribution channels raise questions about this very foundational principle of regulation. Consequently, issues of ownership diversity and localism in particular need to be reconsidered.

In the wake of the deregulation at the heart of the Telecommunications Act of 1996, which led to the unprecedented growth of media conglomerates such as Clear Channel, there has been a renewed concern over ownership and diversity, manifesting itself in the ongoing struggle between faith in the self-regulatory functions of the market and the felt need to guard the public interest against special interests. Perhaps the most contentious round of this struggle since 1996 took place in 2003, when Michael Powell, then chairman of the FCC, proposed loosening restrictions on television station ownership to allow individual corporations to own stations covering 45 percent of the country, up from a previous cap of 35 percent. As Robert McChesney and others have detailed, this proposal was met with unprecedented dissent from the public—the FCC received over 2 million comments against loosening caps. Eventually, Congress altered the proposed rule changes to raise the cap to only 39 percent and passed the measure as part of an omnibus appropriations bill that included a range of spending initiatives popular with both the president and members of Congress.[39]

Large media corporations made two arguments when defending looser ownership restrictions. First, following the logic of free market economics, they suggested that monopolies are more likely to create diversity because they can afford to offer a range of differentiated programming for small segments of an overall audience.[40] The US media system is an example of a limited competition model that encourages a range of competitors in the marketplace. But the cost of entry is so high that smaller competitors are not encouraged to take big risks, relying instead on only slightly differentiated products. Large conglomerates, the argument goes, can reduce risk by spreading it across their vast holdings, making truly diverse programming less risky overall. The second argument they made was that the rules governing ownership were increasingly irrelevant in the new media climate and that adhering to them would jeopardize the continued existence of free over-the-air television by threatening the profit margins of the companies (networks) who were in the business of broadcasting. As Mel Karmazin, then Chairman of Viacom (owner of CBS), explained at a Senate Commerce Committee hearing:

> It is utterly unsupportable and unrealistic that broadcasters should be handcuffed in their attempts to compete for consumers at a time when Americans are bombarded with media choices via technologies never dreamed of even a decade ago, much less 60 years ago when some of these rules were first adopted.[41]

For the proponents of consolidation, the post-network era is defined by changes in the media economy that current regulatory strictures cannot adequately address. Especially since Reagan's appointment of Mark Fowler as the chairman of the FCC in 1981, Congress and the FCC have consistently relaxed ownership rules in ways that have favored consolidation and the growth of large media conglomerates. At the same time, the power and reach of these conglomerates continues to be an area of public and political concern, leading to renewed calls for regulation on the grounds of the public interest mandate put in place for broadcasters early in the twentieth century.

Media reformers and critics of deregulation argue that consolidation leads to uniformity and a move away from localism: the bedrock of the public interest mandate. For this group, the control that advertisers exert over media companies necessarily encourages content that is more conservative in nature and geared to the broadest possible audience. More importantly, these critics argue that consolidation threatens to shut down local and independent voices. As Kevin Howley has argued, with regard to a 2007 proposal to ease restrictions on media cross-ownership:

> The FCC's largess comes at an enormous price to the public interest. Media concentration shuts out independent voices, undermines local news operations and eviscerates investigative journalism in local communities. Rather than foster competition and media diversity, the FCC's ruling simply allows big media to get even bigger.[42]

As an example of what consolidation means for local news, critics can point to the specter of central casting, in which large station owners such as NBCU, Belo, Hearst–Argyle, and Sinclair consolidate their programming and production operations in a central location and feed them to individual stations. Although these companies and others have long centralized their master control operations and graphics generation, many station owners are considering centralizing news and weather operations as well. Such a move would call into question the very definition of localism: "Is it locally originated programming? News and sports? A physical building? Or is it the subtle influence of the shared destiny of a station and the community it serves?"[43]

Localism as a regulatory category has been ill-defined historically. As Robert Horwitz argues, while localism has always been a foundational principle of broadcast regulation, it has also been "a profoundly ambiguous concept" and one which the FRC and later the FCC have been reluctant to define with any "substantive criteria that might secure ... broad public interest goals."[44] One reason for this ambiguity may lie in the contradictory nature of the broadcast industry in the United States itself: it is at once a public trust and a commercial business. The incentives toward localism are not so much born out of a sense of duty or obligation to the betterment of the community but, rather, come from the market which might well reward a broadcaster that seems connected to the particular community of which it is a part. A second reason for the ambiguity may come from the mostly arbitrary construction of local markets, based on technical elements rather than an social or cultural factors. Finally, a third reason for this ambiguity may stem from uncertainty about whether localism is a matter of actual content or physical presence. Each of these potential factors has only become more pronounced in the post-network era and requires us to revisit the very idea of localism.

As an example of the dynamics that drive localism in the post-network era, consider this example from radio (which could easily describe a potential future for local television). The passing of the Telecommunications Act of 1996 opened the door for Clear Channel to purchase over 1,200 radio stations—a development widely hailed as evidence of the erosion of localism by deregulation. Yet, one result of this contraction has been the development of hyper-local radio formats in suburban markets, such as those operated by NextMedia, which operates stations in suburbs of Chicago and Dallas. The hyper-local ethos of these stations is evident in a statement on the website of NextMedia's Star 105.5 in Crystal Lake, Illinois:

> Star 105.5 is located at 8800 US Highway 14 in Crystal Lake, in the same vicinity as McHenry County College. Turn at the stoplight directly in front of the college, follow the road around and we are just behind the Fire Training Facility.

In this case, however, hyper-localism seems to be a matter of merely mapping the station into the local geography rather than a substantive shift in programming priorities. In fact, hyper-localism is less a regulatory imperative than a market imperative: a way of setting oneself apart in an increasingly crowded field. With this in mind, Tom Rosenstiel, director of the Project for Excellence in Journalism, suggests that hyper-localism may not be the panacea that media reformers would hope for, and that it may, in fact, represent a negative contraction in the operation of a station rather than an expansion of its most valuable asset—its connection to the local market:

> The current thinking, hyper-localism, seems problematic. In an era of globalism, how can you suggest that the L.A. or Boston market does not need its own specialized foreign reporting that informs the local economy, the local culture and more, in a way that is different than what generic wires would cover?[45]

The problem of localism is that it is not clearly beneficial given the rather narrow geographic parameters within which it has been defined. It remains to be seen whether hyper-localism takes shape via low-power television stations broadcasting to very specific geographic sectors or through mobile and Internet-based distribution that allows users to pick and choose content on demand. But, in the era of "big media getting bigger," localism has become a key component of both political and commercial interests and will shape the operations and image of local stations to an even greater extent than before.

In an era of rapid institutional change—both structural and technological—it is notable that the basic regulatory structures (with a few exceptions) have not been significantly altered. Proponents of consolidation may be right in principle that the rules devised 60 years ago no longer adequately address the current media landscape. It may also be true, however, that core ideas about localism and diversity need to be reconsidered. As Christopher Anderson and Michael Curtin note: "each passing day makes it more difficult to clarify the media's role in defining the social meaning of place, particularly as individuals increasingly inhabit a global cultural economy."[46] In a post-network environment in which local stations are operated predominantly by non-local interests, station owners, regulators, and critics alike will need to consider carefully what it means to be local and what steps will be taken to ensure that local interests survive in this new environment. Because localism has never been adequately defined, it can be claimed to function well under one of several conditions, including deregulation of ownership caps. Anderson and Curtin raise concerns when they note: "As government regulators formulate policy in response to … technical innovations, particularly in a political climate that favors deregulation, it will be revealing to see whether a commitment to localism survives even in principle."[47] In the post-network era, there may be an opportunity to redefine local interests and *re*-regulate for them so long as there remains a commitment to localism in principle.

Conclusion

It may be time at least to begin thinking about local television in a truly post-network sense: as stations existing without affiliation in local markets that may not be able to accommodate multiple competitors for essentially the same product. Structural and technological developments have significantly shifted the media landscape to the degree that old business models simply don't make sense. But this is a transitional moment driven by transitional logic in which older models have not entirely given way to new and emergent trends. The dynamics of local television are defined by a range of contradictions and ambiguities: between economic expansion and contraction, between national (even global) reach and local commitment, and between autonomy and centralization. These dynamics are indicative of the larger culture industries as well and help define what we are calling the "post-network era." As the previous sections have illustrated, the future prospects for local television stations will depend on an interconnected web of economic, technological, regulatory, and creative factors. The task of placing stations within the context of the post-network era requires that we confront all of these issues.

From an economic perspective, local stations find themselves in positions of both strength and weakness—value and expendability—in dealing with larger media organizations. With regard to cable and satellite delivery services, local signals represent a valuable asset, and the most recent round of retransmission negotiations with cable operators have highlighted the new-found leverage that local stations possess. This leverage is due at least in part to the number and variety of stations that one ownership group could pull from a cable system should negotiations stall, as well as to the increased competition for signals that satellite companies represent. At the same time, however, local stations are in a position of increasing weakness in relationship to their network partners. Battles over clearance agreements and reduced (even reversed) compensation have underscored the relative hegemony that the networks enjoy with regard to the operation of local stations. Even local news—the one remaining mainstay of local production and community involvement—is increasingly dictated by the demands of the networks, which insist on a certain number of hours of local news to support their own offerings.[48] Stations are actually expanding their news production because local news is such a dependable profit center for local stations and the key element of their leverage in negotiations with cable and satellite operators. But, should their overall audience continue to shrink or migrate to on-demand platforms, a new point of tension between the stations and the networks is likely to emerge and threaten the very relationship upon which the stations' current existence and operation is predicated.

Technological developments such as Internet portals and mobile distribution capabilities highlight a key tension between expanded (national or even global) reach and the commitment to local communities. Local stations can now make use of their own websites and/or more expansive video-sharing sites such as YouTube to distribute news packages. These new locations open up new marketing and revenue-generating possibilities for stations whose traditional advertising sales are in a slow but steady decline. From a central location, local stations can now reach an expanded audience for whom physical proximity no longer necessarily defines community. At the same time, however, these same technologies of interconnection and expansion allow large station owners to centralize the operations of multiple stations. The same news packages, weather updates, and sports stories that drive local news can be made more cost efficient and profitable through off-site production. While this practice is not widespread, and would certainly generate controversy in more critical and community-minded circles, the corporate logic of cost-cutting through expedience and efficiency make it a rational—if politically or ethically questionable—possibility. Like a number of FOX affiliates currently programmed by the staff of a sister station/competitor in the same market, the drive toward centralization and profit often outstrips critical notions about the democratic function of diversity and competition in the marketplace of ideas. As an employee of WTWO in Terre Haute (which also owns and operates WFXW, the FOX affiliate in the same market) explained: the FOX affiliate in Terre Haute consists of a transmitter and a piece of paper. While the FOX station still exists technically, that existence is predicated solely on the continued existence of the traditional broadcast–affiliate model. Should FOX ever decide to bypass local broadcasting, WFXW and other stations like it would disappear from the air without ceremony.

Finally, renewed critical and regulatory interest in localism demands that stations and regulators alike reconsider what it means to be local and how a commitment to localism will be measured. As discussed earlier, historically localism has been an ambiguous concept at best, based more on twin measures of physical proximity and local production hours. But as both of these factors come into question in the wake of the technologies of centralization, new ways of thinking about localism and the marketplace are required. Some scholars suggest that media companies should be "held accountable for the cultural environment in the same way that companies need to be regulated in terms of the natural environment."[49] Although such attention to the cultural environment sounds promising, it runs headlong into the "culture wars" that have animated the polarized politics of the last three decades. Additionally, the expanded reach and specific targeting allowed by new distribution technologies necessarily call into question the viability of any traditional sense of programming for a community. What this means for local stations is that they find themselves caught between old ways of imagining community and new ways of reaching audiences. In this environment of deregulation, it is possible that legislative concerns about community-based localism could disappear, replaced by the enthusiasm for new, more personalized distribution technologies. In the event of a new sense of local as personal, local stations could find themselves too committed—financially and culturally—to old ways of doing business. Whatever transpires in the realm of regulation, it will likely lead to an end of local stations as we have known them.

The demise of local affiliation is not a fait accompli, though. The government-mandated switch to digital broadcasting has forced all stations to invest millions in upgrading to digital transmission equipment, and since many of these stations are owned and operated by the networks themselves, there is a financial incentive to continue the basic relationship into the foreseeable future. Predicting this future for local stations is inherently difficult. Unlike the major networks, which will likely respond in similar fashions to the revolutionary changes that confront the media industries, local television stations are too varied in function, resources, and ownership structure for there to be consensus or even consistency. What is certain, however, is that local stations as we have known them for 60 years are going through something of a revolution themselves. And, as with any revolution, the transitional moment is fraught with contradictions and ambiguities.

Local stations will continue to exist, but their fate is increasingly separate from that of the networks with which most of them have been connected, and their function, structure, and numbers may in time look nothing like they do today. I have tried to illuminate the major factors that are driving these changes in order not only to explain what is happening, but to suggest ideas about why it might matter. In a cultural environment in which information about major global and local events comes to us via any number of increasingly interconnected media outlets, the health of a vital democracy is dependent on the existence of independent voices. The current state of affairs—economic, technological, and regulatory—that defines the post-network era represents both a promise and a threat. Local stations could stand at the vanguard of a renewed and vital media system, freed from the constraints of major commercial interests of the networks. But, at the same time, the acceleration of new technologies, combined with the momentum of historical ways of doing business and of regulating, threaten to give us more and more of the same, even if it looks like a brand new world.

Notes

1 Once technological innovations are accounted for, most histories of network television focus on the economics, regulation, programming, and cultural impact of the major networks. See Eric Barnouw, *Tube of Plenty: The Evolution of American Television* (2nd ed., New York: Oxford University Press, 1990); J. Fred MacDonald, *One Nation Under Television: The Rise and Decline of Network TV* (New York: Pantheon Books, 1990); Douglas Gomery, *A History of Broadcasting in the United States* (Malden, MA: Blackwell, 2008).

2 Univision, the top-rated Spanish-language broadcast network in the United States, accounted for twelve of those programs rated between 56 and 100. The Television Bureau of Advertising, http://www.tvb.org/nav/build_frameset.asp?url=/rcentral/index.asp (accessed July 2, 2008).

3 Ibid.

4 Ibid.

5 According to research conducted and compiled by the Project for Excellence in Journalism, the top 25 markets accounted for over half of total station revenue in 2007: http://www.stateofthenewsmedia.com/2008/narrative_localtv_economics.php?cat=2&media=8 (accessed June 25, 2008).

6 Joe Schlosser, "Fox is 'King of the Hill,'" *Broadcasting and Cable*, May 4, 1998, http://www.highbeam.com/doc/1G1-20581062.html (accessed August 7, 2008); Melissa Grego, "Bochco Red Hot over 'Blue,'" *Broadcasting and Cable*, September 20, 1999, http://www.highbeam.com/doc/1G1-56067296.html (accessed August 7, 2008); Mara Einstein, *Media Diversity: Economics, Ownership, and the FCC* (Mahwah, NJ: Lawrence Erlbaum Associates, 2004), 53–4.

7 Klinenberg's argument is centrally concerned with the problem of ownership and its impact upon democracy. Eric Klinenberg, *Fighting for Air: The Battle to Control America's Media* (New York: Metropolitan Books, 2007), 114. See also Robert McChesney, *The Problem of the Media: U.S. Communication Politics in the Twenty-First Century* (New York: Monthly Review Press, 2004).

8 Tim Brooks and Earl Marsh, *The Complete Directory to Prime Time Network and Cable TV Shows, 1946-Present* (7th ed., New York: Ballantine Books, 1999), 1250.

9 Michael Curtin, "On Edge: Culture Industries in the Neo-Network Era," in *Making and Selling Culture*, ed. Richard Ohmann *et al.* (Hanover, NH: Wesleyan University Press, 1996), 197.

10 Charting ten stations across one week, and including a range of market sizes and network affiliations, I found that the typical network affiliate schedule contained an average of 13.5 hours of network programming.

11 Bill Carter, "CBS Discloses Plan to Begin Charging Fees to its Affiliates," *New York Times*, May 31, 1992.

12 Jenny Hontz, "Webs Remodeling Station Compensation Plans," *Variety*, January 18, 1999, http://www.highbeam.com/doc/1G1-53706164.html (accessed July 8, 2008). See also Howard J. Blumenthal and Oliver R. Goodenough, *This Business of Television: The Standard Guide to the Television Industry* (3rd ed., New York: Billboard Books, 2006), 9.

13 Ken Auletta, *Three Blind Mice* (New York: Random House, 1991), 263.

14 Carter, "CBS Discloses Plan," A1.

15 Ibid.

16 Katy Bachman, "Nets vs. Affiliates Battles Continue," *MediaWeek*, April 8, 2002, http://www.allbusiness.com/services/business-services-miscellaneous-business/4821960–1.html (accessed March 1, 2007).

17 Michele Greppi, "NBC Affiliates Prepare to Swallow Bitter Reverse Compensation Pill," *TVWeek*, July 13, 2008, http://www.tvweek.com/news/2008/07/nbc_affiliates_prepare_to_swal.php (accessed July 15, 2008).

18 "Fox Playing Hardball," *Television Digest With Consumer Electronics*, November 8, 1999, http://findarticles.com/p/articles/mi_m3169/is_/ai_n27553787 (accessed July 2, 2008).

19 Jay Sherman, "Chancey Indie Road Ahead," *Television Week*, January 30, 2006, http://www.highbeam.com/doc/1G1-141651671.html (accessed July 3, 2008).

20 Bachman, "Nets vs. Affiliates."

21 For a useful history of these beginnings of cable, see Michele Hilmes, *Hollywood and Broadcasting: From Radio to Cable* (Urbana-Champaign: University of Illinois Press, 1990).

22 Michael Kassel, "Must-Carry Rules," *Encyclopedia of Television*, ed. Horace Newcomb (2nd ed., New York: Fitzroy Dearborn, 2004), 1579.

23 Satellite penetration reached 27.7 percent of television households in May 2008, according to the Television Bureau of Advertising and Nielsen data: http://www.tvb.org/nav/build_frameset.aspx.

24 Linda Moss, "MSOs Draw Line at Cash for Carriage," *Multichannel News*, January 3, 2005, http://www.multichannel.com/article/CA490771.html?industryid=47201 (accessed June 16, 2008).
25 Because of political advertising dollars that help raise totals in even years, comparisons are more useful between odd years.
26 Project for Excellence in Journalism, "State of the News Media 2008," http://www.stateofthenewsmedia.com/2008/narrative_localtv_audience.php?cat=1&media=8 (accessed July 16, 2008).
27 "Eisner Questions Need for Affiliates," *Television Digest with Consumer Electronics*, October 5, 1998, http://findarticles.com/p/articles/mi_m3169/is_n40_v38/ai_21191675 (accessed July 1, 2008).
28 Michele Grepi, "Affils See Red over iTunes Pact," *Television Week*, October 17, 2005, http://www.highbeam.com/doc/1G1-137823451.html (accessed September 2, 2008).
29 See Brooks Barnes, "As TV Networks Use Web, Affiliates Seek Piece of the Action," *Wall Street Journal*, February 1, 2006, B1; "ABC Affils Wade into Web Stream," *Daily Variety*, May 8, 2006, 6; "Smorgasbord of TV on Network Sites," *Television Week*, October 9, 2006, 15; Ellen Lee, "Fox Will Offer Shows Online," *San Francisco Chronicle*, April 15, 2006, C1.
30 Duane Lammers, interviewed by author, Farmersburg, IN, January 7, 2006.
31 Bob Papper, "Web Survey: Connected," *Communicator*, May 2008, http://www.rtnda.org/pages/media_items/web-survey-connected1055.php (accessed June 28, 2008).
32 Ibid.
33 "TV Stations' Online Bet Should Be All In," *TVNewsday*, June 17, 2008, http://www.tvnewsday.com/articles/2008/06/17/daily.2/ (accessed June 17, 2008).
34 Mary Collins, "Opportunities Abound in the Digital Age," *TVNewsday*, August 22, 2008, http://www.tvnewsday.com/articles/2008/08/22/daily.3/ (accessed August 25, 2008).
35 Ryan Nakashima, "Local Station Owners Push Mobile TV," http://www.msnbc.msn.com/id/24116669/ (accessed July 11, 2008).
36 Ibid.
37 Christopher Anderson and Michael Curtin, "Mapping the Ethereal City: Chicago Television, the FCC, and the Politics of Place," *Quarterly Review of Film and Video*, 16, 3–4 (1999), 292–5.
38 For a detailed history of how the public interest was defined historically to favor broad commercial interests over specialized (i.e., religious and educational) broadcasting, see Robert McChesney, *Telecommunications, Mass Media, and Democracy* (New York: Oxford University Press, 1993), 12–37. See also Robert Hilliard and Michael Keith, *The Quieted Voice: The Rise and Demise of Localism in American Radio* (Carbondale: Southern Illinois University Press, 2005); and Robert Britt Horwitz, *The Irony of Regulatory Reform: The Deregulation of American Telecommunications* (New York: Oxford University Press, 1989).
39 McChesney, *The Problem of the Media*, 292–3; Stuart Minor Benjamin, "Evaluating the Federal Communications Commission National Television Ownership Cap: What's Bad for Broadcasting is Good for the Country," *William and Mary Law Review*, 24 (November 2004), 2–3.
40 Einstein, *Medix Diversity*, 3.
41 Al Tompkins, "Ownership Caps and Fuzzy Media Math," http://www.poynter.org/content/content_view.asp?id=35660&sid=14 (accessed June 14, 2008).
42 Kevin Howley, "The FCC's Ruling More Corporate Welfare," *Bloomington Alternative*, December 30, 2007, http://www.bloomingtonalternative.com/node/8910 (accessed August 20, 2008).
43 Joan Van Tassel, "Central Casting: All for One, and One for All," *Broadcasting and Cable*, April 23, 2001, http://www.highbeam.com/doc/1G1-74357675.html (accessed August 20, 2008).
44 Horwitz, *The Irony of Regulatory Reform*, 158.

45 James Rainey, "Media's Focus Narrows, Report Warns," *Los Angeles Times*, March 12, 2007, A-8.

46 Anderson and Curtin, "Mapping the Ethereal City," 302.

47 Ibid.

48 An example from the ABC contract with Hearst–Argyle governing thirteen ABC affiliates in 2006 is indicative of the tension between autonomy and dependence:

> Local News
>
> Stations agree to program Monday–Friday at least one half hour of locally produced news leading into ABC's currently scheduled *GMA* morning and currently scheduled *World News Tonight* evening news programs as well as its late night consistent with the Stations' current practices; subject however, to each Station's responsibility under law to determine its programming and the scheduling of its programming. [A sideletter will provide that this next sentence is effective only upon assignment or transfer of the Stations.] However, any reduction in the current amount of local news lead-ins will result in a reduction of the applicable Station's compensation and/or its preemption basket in an amount to be determined by good faith negotiations of the parties.
>
> http://www.secinfo.com/d11MXs.vcZc.d.htm (accessed August 21, 2008)

49 Einstein, *Media Diversity*, 4.

Chapter 9

Reinventing PBS

Public Television in the Post-Network, Post-Welfare Era

Laurie Ouellette

In 2000, in the wake of plummeting ratings, a drop in viewer donations, waning corporate underwriting, and public funding cuts, the Public Broadcasting Service (PBS) announced a plan to "reinvent" itself. Under the new leadership of former cable executive Pat Mitchell, the interconnective hub of the United States' 354 public television stations spearheaded a two-pronged approach to reinvigorating the system's fading vitality and shrinking economic base. The first strategy was to make public television more entrepreneurial and competitive in a changing cultural marketplace. This has involved streamlining business operations, updating PBS programming, rebranding its image, forging commercial partnerships, and expanding revenue-generating activities across broadcast and new media platforms.[1] The second strategy was to update public television's non-commercial public service mission for the digital era and identify new justifications for public and philanthropic funding. PBS entrusted the private Digital Future Initiative—a self-described Carnegie Commission for the emerging stage of public television—with this agenda, which has yet to be fully realized.[2]

PBS is not the only public broadcaster to succumb to the spirit of reinvention. Across Western capitalist democracies, Georgina Born explains, the "conditions—technological, economic, political, social and cultural" that once fostered public broadcasting have undergone "such a radical transformation that the concept and practice … demand to be reconceived."[3] Born cites the decline of national cultures, increasing individualization, and the flux introduced by digitalization and public sector reform initiatives as factors in the ongoing transformation of European public broadcasting.[4] While these pressures are also evident in the United States, the situation facing PBS is more complex vis-à-vis the British Broadcasting Corporation (BBC) and other national systems. Public television arrived late in the United States, as a corrective (but structurally marginalized) supplement to the commercial model of broadcasting dominant since the 1920s. Created during the expansion of social welfare programs in the 1960s to fill cultural gaps in an over-the-air broadcast schedule dominated by ABC, CBS, and

NBC, it has lost much of its footing as the television marketplace has expanded and the broader socio-political climate of the United States has changed.

This chapter analyzes the reinvention of PBS as a response to the double bind it now faces: *a loss of place* as commercial channels multiply and new media technologies emerge, on one hand, and *a loss of legitimacy* as the current stage of "neoliberalism" rethinks the interventions of the 1960s—such as PBS's creation—and advances a privatized approach to the public good, on the other. Taking public television's makeover as an occasion to historicize the present, I situate its adaptation strategies within changing rationalities of culture and citizenship and show how US public television's development as a solution to presumably "resolved" problems clouds thinking about the future. The old PBS provided a bridge between the commercial network era of broadcasting and the triumphant declaration of choice, freedom, and empowerment surrounding both the post-network telescape and the concomitant transition to a post-welfare society. This legacy is the starting point for evaluating the PBS's attempt to reinvent itself as a business, a brand, and a resource for the new epoch of mass customization and personal responsibility.

Inventing PBS: A Brief History of the Present

My book *Viewers Like You?* situates the cultural formation of US public television within the fissures of US liberalism, an approach to governing in which the rationality of the market is valued over direct state involvement in the social.[5] While European nations were establishing public broadcasting bureaucracies as technologies of education and national citizenship, US policymakers authorized privately owned corporations to operate commercial broadcasting "in the public interest." As television became a mass medium, the limitations of this arrangement became evident to reformers and policymakers. At a time when television ratings were still based on the number of viewers (rather than their demographics) and channels were "scarce," three major broadcast networks competed for the largest audience possible. Programming was approached as bait and filler for the sale of advertising—the business of broadcasting. While commercial television was valued as the engine of the consumer economy, its cultural output was increasingly deemed at odds with the core liberal attributes—individualism, pluralism, self-governance, robust civil society—ascribed to US democracy. High anxieties about television's standardization and homogeneity, proclivity to trivial amusements, and association with a mass audience presumed to be indiscriminate and ill-equipped to carry out the rights and responsibilities of citizenship gained currency in the 1960s, memorialized by FCC Chairman Newton Minow's 1961 "Vast Wasteland" speech.

As concerns about television escalated, powerful corporate philanthropies—especially the Ford and Carnegie foundations—mobilized for an alternative program service, providing another stimulus and some of the seed money for what eventually became PBS. The interventionist approach to social welfare associated with the "Great Society" programs of the 1960s made it possible to rethink television as a technology of public education and citizenship formation more broadly. Federal investment in public education, health, and welfare rose dramatically during the 1960s in response to the demands of the emerging information economy, the Cold War, and percolating social unrest. Subsidies to the arts expanded as well, reflecting a national commitment to aesthetic and cultural values deemed unachievable through market mechanisms. The Great Society era's pressing concern with raising the nation's quality of life, equalizing educational opportunities, and improving people's capacities in all spheres of life also placed a new premium on broadcasting as an instrument for maximizing human capital and instilling standards of excellence. Commercial television's refusal—or inability—to allocate resources to these goals prompted President Lyndon B. Johnson and eventually the US Congress to support the idea of a national non-commercial public broadcasting system. This was seen as an extension of educational television (ETV), a hodgepodge system of non-commercial stations authorized by the FCC in the 1950s to provide formal as well as informal instruction to local communities. With more resources and a broader national mission, US public television would disseminate what Graham Murdock calls "cultural resources for citizenship"—informal education, civic training, cultural enlightenment, opportunities for self-improvement—in tandem with the expanding social and educational services provided by the US government.[6]

Its creators expected PBS, slotted as the "oasis" of the wasteland, to pursue a different mission entirely. Since the commercial networks were criticized for pandering to mass appeal, PBS was called upon to bring "excellence and diversity" to the airwaves, while also recasting the worrisome mass audience as a fragmented array of specialized interest and taste groups. It was regarded as a natural home for well-educated populations with advanced cultural tastes and opinion-leading capacities—the "minorities" deemed least served by the mass appeal-oriented approach to television operating across commercial channels. Predating the arrival of cable television, it developed as the original niche channel, a home for "quality" television with an educational flair, synchronized to what sociologist Pierre Bourdieu calls the cultural dispositions of professional upscale lifestyle clusters.[7] Yet, because PBS relied in part on tax funding, it was also required to uplift disadvantaged populations and facilitate enlightening and purposeful uses of television among all Americans. Expected to protect a small slice of television from the ravages of commercial mass culture, provide programming for sophisticated viewers, and improve the public as a whole, PBS operated as a distinguished and redemptive—but deeply marginalized—niche channel rather than a major broadcaster such as the BBC and other European broadcasters.

While the passage of the 1967 Public Broadcasting Act signaled a reinterpretation of the state's role as financial partner in the mobilization of television as a social and cultural force, a mixed system of funding sources—including Congressional appropriations, viewer donations, philanthropy, and corporate sponsorship, tempered the "socialistic" tendencies ascribed to European public broadcasting and authorized PBS's pioneering role in bringing upscale niche marketing to television. To avoid the centralized power ascribed to state-operated broadcasting as well as the US broadcast corporations, public television itself was also fragmented to the greatest degree possible, composed of the Corporation for Public Broadcasting, hundreds of autonomous former ETV stations, and PBS operating as a new national program service rather than as a traditional broadcast network.

Because PBS was developed as a corrective to commercial television and its imagined audience, its relationship to the people it claims to represent always has been contradictory. On one hand, PBS introduced programming not tied to the imperatives of advertising, a more diversified system of cultural production (with content suppliers outside the entrenched Hollywood system), and early experiments in public outreach and interactivity (from the distribution of educational resources to the staging of mail-in votes on political debates). Yet, PBS has also been positioned outside the sphere of popular television and, with the exception of its children's programs, culturally tied to "influential" professionals with disproportionately high incomes and education levels. Indeed, because of this association, It has been vulnerable to campaigns to downscale the "welfare program for the rich" that began almost immediately, with politically motivated Republicans at the helm. Since the 1990s, the original rationales for public television have lost import across partisan lines. As the commercial telescape transforms and an "entrepreneurial spirit" reshapes the public sector, the notion that PBS requires public subsidy to correct market deficiencies has been difficult to sustain. The US government's retreat from the 1960s—the downscaling of public education and welfare programs and the ethic of personal responsibility for one's fate associated with these reforms—has thrown the Great Society's uplifting aspirations for PBS into flux as well. In short, as television has changed, so too have approaches to welfare and citizenship in the United States. The reinvention of PBS operates at the intersection of these developments.

Doing Good while Doing Well: The New Business of PBS

PBS was flagging at the millennium: ratings were down 23 percent, memberships were on the decline, major underwriters were leaving, and many stations were struggling to stay afloat financially. With myriad commercial cable venues, from CNN to the Discovery Channel, providing similar—but less stuffy—versions of culture, information, and education, the oasis of the wasteland had been losing its distinction for some time. The deepening of the crisis prompted the appointment of Pat Mitchell as a new president/CEO who promised to transform PBS into a more compelling option for corporate sponsors and television viewers awash in a dizzying array of choices. Approaching the reinvention of PBS as a business challenge rather than a cultural policy issue, Mitchell, former CNN executive and head of Time Inc. Television, introduced a new strategic plan in 2001 that reverberates to this day. Similar to the reform of many public institutions, from prisons to the postal system, revitalizing PBS has involved embracing market-derived principles—including efficiency, competition, and entrepreneurialism.[8] Its mission to improve television (and its viewers) has been folded into these principles as the focus has shifted to improving the economic health of PBS, with the market as the elixir. PBS executives operating under the slogan "Doing good while doing well" began moving in this direction in the late 1990s; Mitchell's arrival intensified the process and signaled a changing of the guard.[9]

US public television has always been a public–private partnership, dependent on the goodwill of corporate underwriters as well as tax dollars. What has changed is that the commercial sector has been embraced as an operational model as well as a funding source. This means that PBS, while still non-profit and funded in part by public funds, is being run more like a competitive business than a cultural welfare program. Perhaps the most obvious sign is an intensification of marketing and promotional activities. While program interruptions are still forbidden, brief underwriting credits have morphed into 15- or 30-second commercials—a change introduced to combat the perception that corporate sponsors get more for their money with cable. Integrated marketing deals are another trend, from the Liberty Mutual Insurance Company's visibility at tapings of the PBS program *Antiques Roadshow*, to characters from its children's program *Fetch!* appearing in Macy's in-store promotions and Arby's kids' meals.[10] Commercial sponsorship has also migrated to PBS online, where banner ads and sponsored links provide "added value" to sponsors and new revenue streams unencumbered by FCC regulation.[11] PBS has entered into new commercial partnerships that allow advertising, including the PBS Kids Sprout channel, a co-venture with Comcast cable. And, it has become more aggressive about its own fundraising, merchandising, and branding activities. While critics have rightly protested these escalating "violations" of public television's non-commercial role, it is too simplistic to suggest that PBS has been ruined by market logic.[12] As Born reminds us, institutional change is messy—particularly with PBS, a precarious institution whose "reform" brings the cultural rationalities of the past into conflict with the enterprising demands of the present.

While PBS was created as a non-commercial corrective to broadcast television, it has increasingly approached the major networks (as well as cable competitors) as templates for reinvention. As John Caldwell has shown, the conglomerates that own the major broadcast networks have developed an arsenal of strategies for ensuring the centrality of television—and television networks—in a context of high anxiety triggered by industrial and technological change.[13] In addition to ratings stunts and intensified flow techniques, these moves include repurposing content across corporate holdings, rebranding campaigns, and promotional activities to educate viewers and affiliates about the "benefits of national network affiliation."[14] While PBS operates on the fringes of the industrial context charted by Caldwell, it has adopted similar tactics in tension with its positioning as "better" television untainted by market pressures and standardizing industrial processes. As with the commercial networks, the new PBS pursues its own survival over earlier cultural ideals—including its historical role as a curative technology of liberal pluralism and enlightened citizenship.

PBS's embrace of a culture industry model is not inherently bad. As Justin Lewis points out, industrial techniques developed for commercial purposes (such as marketing and audience cultivation) can be used to facilitate progressive social and cultural goals as well as to reduce the elitism associated with much public cultural subsidy.[15] PBS, however, has exploited the latest business logic not for the policy-driven reasons discussed by Lewis but to ensure its capacity to survive—and thrive—in a competitive media environment. For example, to ensure more cultural diversity and localism than was apparent in commercial network broadcasting of the 1960s, PBS was designed to coordinate a national program schedule from a dispersed collection of content suppliers, including local stations, independent producers, exporters, and regional production centers. Stations—unlike commercial affiliates—were allowed to decide which PBS programs to broadcast and when to schedule them. Under revision since the mid-1990s, this flexible approach to production and distribution has emerged as a perceived barrier to "efficiency" as a cure for institutional malaise. Besides prompting the downscaling of PBS headquarters and discussions of station mergers and consolidations, the pursuit of efficiency has come to bear on the public television production process, with PBS urging producers to "streamline" operations, avoid redundancies, and coordinate around "central principles" to maximize fundraising efforts.[16] PBS also developed a new "declaration of interdependence" to unify stations in support of a stricter common carriage policy, on the grounds that a unified national prime-time schedule—in which stations follow the lead of PBS programmers rather than devising their own schedules based on local preferences or needs—would improve the system's economics as a whole.[17] In sync with technologies of audit and "outcomes assessment" being implemented across the public sector, PBS tempered the loss of local control inherent to common carriage by inviting stations to "measure" the results of its service through ratings increases and membership gains. While some have resisted this logic, the move toward a more centrally coordinated infrastructure is a case where the enterprising spirit increasingly overrides earlier assumptions about public television's curative role in liberal democracy. Compromising the decentralization prized by broadcast reformers hoping to combat commercial television's role in the massification of society, PBS has repositioned itself as a viable national network for business reasons—including the pursuit of corporate sponsors and the maximization of its own branding and self-marketing campaigns.

Likewise, the quest for higher ratings not only assumes competition with other television channels—which PBS has historically avoided—it also runs counter to an earlier institutional ethic of selective and purposeful viewing adopted to correct the ritualized and indiscriminate habits ascribed to the mass audience. To foster more specialization and individual choice in television, PBS encouraged potential viewers to watch specific programs based on self-defined tastes, interests, and needs. While some scheduling tactics were apparent (e.g., children's programs were shown during the day), prime time was not designed to "capture" an audience in the manner pursued by commercial networks. Shows did not flow seamlessly from one to the next and, because inertia was discouraged, there were no teasers to keep the dial tuned to PBS. On one hand, this model favored an educated disposition toward appointment viewing, reifying once again the "natural" match between the upper-middle class and PBS. Yet, it also challenged the commercial management of audiences and anticipated the customized viewing and scheduling practices associated with today's new recording, distribution, and viewing technologies. In pursuit of viewers and dollars, PBS has moved away from its earlier ethic of selective and purposeful engagement toward intensified sequencing and flow techniques pioneered during the network stage of commercial television, including scheduling strategic programming blocks designed to move viewers from one ratings time sample to the next on a given evening (e.g., history night, science night, public affairs night). Promotional "bridges" between programs have also been designed to prevent viewers from leaving the PBS audience during local station breaks.[18] While PBS has also made some content available for viewing on demand and online (as will be discussed later), this has occurred alongside a move to control the choices and habits of television viewers for business purposes.

The turn to rebranding as a solution to cultural stagnancy also takes its cue from the culture industry. The perception that PBS is "dull and boring," combined with the legacy of its Great Society mission to uplift and enlighten television viewers, was considered a factor in PBS's loss of ratings. As one trade journalist stated, "Viewers often equate PBS programming with spinach. It may be good for them, but they don't want to eat it."[19] The new strategic plan authorized a makeover, but, as the original niche channel, PBS has been unwilling to shed its ties to well-educated and upscale lifestyle clusters. As the bread and butter of public television, influential viewers with professional incomes and credentials have always been sold to corporate sponsors and "desperately sought" as financial contributors.[20] During its formative period, PBS presented a unique option for TV viewers in search of distinctive television and corporations hoping to reach the "class within the mass." Viewer flight to cable competitors has eroded the synergy between the goals of cultural reformers and the demands of fundraising. The aging of the PBS audience (particularly vis-à-vis cable viewers) and the drop of the average PBS rating to 1.8 percent of the available audience, as reported in 2002, contributed to a feeling that PBS had lost its "relevance."[21] While, because of its educational ethos and reliance on niche marketing, PBS will never be a mass network, it has attempted to cultivate a more compelling brand identity. In addition to creating its own branding department, it hired the Fallon advertising agency to overcome a previous era's understanding of public television and encourage more—and particularly younger—viewers to see PBS as an expression of their identity and lifestyle. With goldfish jumping out of bowls and chamber musicians smashing their instruments, the new promotional spots signified a break from the past to situate the new PBS as a hipper, more fun, and contemporary brand of culture, information, and education.[22]

Investment in the PBS brand has spilled over into public television's approach to multicasting, where it supports the turn to cost-efficient repurposing strategies rather than the use of new channels to diversify programming. As Caldwell argues, repurposing developed in the context of corporate media consolidation as a technique for maximizing profit by spreading the costs of production across corporate holdings. When the same content appears on NBC and CNBC, it provides the basis for another round of advertising sales. Pioneers in the transition to digital broadcasting, many PBS stations have been multicasting on two to four digital channels, available to all television viewers with digital televisions since 2006. But, to fill the new platforms, the stations have turned mainly to prepackaged, sub-branded PBS channels, including PBS World (public affairs), Create (how-to and lifestyle), and PBS Kids. While presented as the expansion of choice, channels merely repurpose programming from the existing PBS library for more precisely calibrated taste and lifestyle niches. Funded in part by the Corporation for Public Broadcasting, the digital channels create new revenue streams for public television while also bolstering brand loyalty and merchandising opportunities through their tie-in websites. New rounds of corporate underwriting are solicited for the multicast channels, and additional exposure for original sponsors is being leveraged in package deals.[23] While a few stations also operate local digital channels, most lack the resources to use multicasting for anything more than the fragmentation of the PBS brand for specialized audience streams and the ability to "co-brand" by inserting local station IDs and announcements.[24]

Repurposing and branding are also co-principles in PBS's commercial collaborations, including V-me, a new Spanish-language channel created by PBS station WNET (New York), and the Baeza Group, a venture capital firm, which is being offered as another digital multicasting option to PBS stations. Conceived as high-end cultural and educational programming for an upscale Latino market uninterested in the popular genres (such as telenovelas) found on other Spanish-language channels, V-me simultaneously trades on and extends the distinguished PBS brand: in addition to some new programming backed by corporate underwriters, it features a heady dose of repurposed PBS programs dubbed in Spanish.[25] While V-me is being pitched as an upscale channel, in the typical way that public television has always conflated diversity with upscale cultural capital, it is technically available to all with digital tuners within the vicinities of subscribing stations. However, other ventures, such as PBS Kids Sprout, a joint venture of Comcast, HIT Entertainment, PBS, and Sesame Workshop (creators of *Sesame Street*), unapologetically compromise the promise of universal service historically associated with public television. In this case, PBS supplies the brand value to what is being billed as a public–private partnership, and many children's programs associated with PBS appear on the basic cable channel, which requires a cable subscription and which allows full-length advertisements before and between programs.[26] As a move toward the privatization of public service, PBS Kids Sprout exemplifies some of the strategies and recommendations of the Digital Future commission discussed later in the chapter.

Programming, Interactivity and the Art of Effects

The reinvention of PBS has bolstered long-standing concerns about the "creeping commercialism" of US public broadcasting. For many critics, the issue is not only the increased privatization of public service and public culture in general, but the extent to which consumerism is replacing citizenship as a normative mode of address. Such complaints often pit taken-for-granted categories—citizenship (good) and consumerism (bad)—against each other, without attending to the complexities of what each entails within the changing contours of public broadcasting. In fact, the imperative to "sell" PBS as a network and a brand conflates these binaries, as market-driven promises of pleasure and participation are increasingly folded into residual and emergent thinking about "good citizenship" in the United States.[27]

In its formative years, PBS pursued a dual mission of mobilizing "opinion leaders" presumed to play a leading role in liberal governance, while also transforming television viewers at large into enlightened citizens.[28] This approach valorized a hierarchical democracy, with educated white men with legitimated cultural capital at the top, and a feminized working class presumably addicted to mass amusements at the bottom. To the extent that PBS programs were closely bound to university-sanctioned tastes and aesthetic dispositions, they serviced mainly the top tier of citizenship so defined. However, PBS also promised to diffuse valorized knowledge, skills, and sensibilities as part of its mission to provide cultural resources for citizenship in tandem with Great Society goals.

These dual agendas formed the basis of PBS's engagement in what Ien Ang (following Michel Foucault) calls the "art of effects," or an early public television supporter called the "great power to affect the behavior of individuals and groups."[29] Drawing implicitly on Foucault's theory of governmentality, which emphasizes the extent to which liberal democracies govern through dispersed techniques for guiding and shaping conduct, Ang challenged the assumption that the effects pursued by public broadcasting (such as elevated taste and good citizenship formation) are inherently progressive. While public service aims are different from a commercial imperative to "win" an audience for advertisers, they are not necessarily less controlling: there's a fine line between enlightenment and social management when television's non-commercial purpose is to reform the public in order that it "better perform its democratic rights and duties," Ang contends. Taking the BBC as an example, she theorized the refusal of audiences to cooperate with top-down public service aims once commercial channels became available as evidence of resistance or even unruliness. The loss of PBS viewers can be similarly interpreted, with important caveats.[30] To win them back has involved tempering the art of effects and folding residual public service aims into a consumer orientation. Not as simple as "selling out," the imperative to sell PBS as a brand has opened up new possibilities and closed down others.

PBS's new strategic plan reframed the earlier two-pronged mission of mobilizing and uplifting citizens as the need to transform "social capitalists" who demonstrate good citizenship by voting, volunteering, attending museums and cultural performances, and participating in their communities into PBS viewers and donors.[31] The term "social capitalist" indicates the fusion of good citizenship with enterprising activity—on the part of individuals as well as PBS—which legitimated its concern with winning upscale viewers in part by idealizing them as good citizens. Because citizenship is commodified, PBS must also conceptualize its imagined audience differently. Two changes are apparent: first, populations who do not possess the advanced social capital ascribed to positive citizenship no longer figure as prominently as subjects to be enlightened, on account of their lack of market value; second, PBS has had to speculate on and address the wants, as well as the needs, of prospective viewers. This has involved minimizing association with an anti-pleasure principle by updating PBS programming and fostering viewer involvement and fan activity. PBS has had to broaden its definition of good television slightly to encompass some dimensions of popular culture coded as trivial, lowbrow, and feminine. However, these gains have materialized within the constraints of niche marketing and in conjunction with new strategies for channeling the energy and labor of brand loyalists into the campaign to restore the health and vitality of PBS.

Short on resources and wishing to retain the still valuable dimensions of its educational legacy, executives have not radically changed prime-time PBS programming. However, some notable tweaks and updates are apparent. Some programs have been repackaged into user-friendly formats, such as *American Stories* (2001), which repurposes material from Ken Burns's documentaries into hour-long formats. PBS has also experimented with forays into popular entertainment: it picked up two seasons of *American Family* (2002–3), an episodic, star-filled drama about a Latino family that failed as a CBS pilot. More recently, PBS signed on for a trial run of *Click and Clack's As the Wrench Turns* (2008), an animated sitcom based on National Public Radio's successful *Car Talk* franchise. The more pervasive trend has involved modernizing iconographic PBS genres and developing new twists on others. The documentary series *Frontier House* (2002) drew from the conventions of popular reality television by putting "modern families in 1883 clothes, wagons, and cabins" and having them "prepare for a Montana winter with the tools of the age" while the cameras rolled (*Texas Ranch House* followed a similar format). *History Detectives* (2003–) puts researchers, forensic experts and appraisers to work answering everyday historical questions about old objects, popular culture, and family genealogy. *Wired Science* (2007–), a partnership with *Wired* magazine, also emphasizes the personalized dimensions of science and technology and clips along at a much faster pace than usual for PBS. *Travis Smiley* (2004–) combines interviews with authors, newsmakers, and celebrities in the late-night slot, blurring public television's long commitment to hierarchical categories of hard and soft news. The *Masterpiece Theater* series (1971–), a PBS staple for more than 30 years, was renamed *Masterpiece* when long-time sponsor ExxonMobil left. Gillian Anderson of the *X-Files* fame was brought in as the new host, the opening visuals and graphics were updated (no more tweed and overstuffed libraries), and both more US productions and a contemporary and mystery version of the franchise were added.[32] "How-to" lifestyle programming on topics such as finances, skin care, and health has proliferated as well. PBS also intended to develop more innovative and interactive public affairs programs to court an audience of "social capitalists," but was unable to secure funding for reasons discussed later in the chapter.

While PBS has tried to shed its schoolmarm image, it has not strayed *too* far from its upscale professional niche. As market strategy, the makeover is less about making PBS accountable to broad popular demand than it is about catching up with the tastes and desires of upper-middle-class lifestyle clusters and positioning PBS more securely within them. If the PBS audience is now imagined to prefer lighter forms of culture and education, it is no more diverse—particularly in terms of the income, education, and cultural capital—than in previous eras. As proudly stated on the PBS website, PBS viewers are still a "distinctive" slice of the population, more apt to invest in stocks and bonds, own vacation homes, attend museums, and possess postgraduate degrees. The new PBS has not courted new constituencies who "don't count" on business spreadsheets or whose tastes and desires might jeopardize its claim to provide a "quality environment for a corporation's brand recognition."[33] Whatever racial diversity has materialized from the new operating rationale has tended to skew to high socio-economic brackets, with *Travis Smiley* attracting the largest black audience *and the most affluent audience* of any PBS program, for example. Despite the relative success of some programs, the attempt to increase PBS's overall ratings has failed. Perhaps anticipating the futility of such a goal in an era of hundreds of channels, PBS executives hoping for numerical gains were simultaneously developing an alternative "points of impact" metric for measuring the value of PBS programming. This has not meant evaluating programming according to non-market goals, along the lines of the BBC and other European public broadcasters who are mandated to serve diverse constituencies and facilitate social goals. Rather, under the new metric the corporate "return on investment" for sponsoring a PBS program is defined as a "combination of ratings, reviews and Internet activity."[34] Because reviewers, viewers, and individuals who seek out PBS on the Internet are assumed to share the educated and upscale attributes desired by corporate underwriters, this system validates (rather than alleviates) US public television's historical association with a small and selective constituency. At the same time, it assigns market value to the particularly active and loyal relationship between PBS and a small but influential audience base, as evidenced by traditional reviews as well as new technologies. While PBS has always used available technologies to "activate" viewers, the point of impact metric has institutionalized audience interactivity as a valuable commodity.

PBS has always encouraged interactivity as a part of its mission to foster purposeful and redemptive uses of television. ETV set the stage by approaching television as a stimulus for "legitimate" activities: reading, painting, studying, discussing, voting. Study guides and resources were often available to casual learners as well as to students taking televised courses for "credit" at local educational institutions. While less overtly instructional, PBS similarly positioned its programs as catalysts to action, whether reading a book, taking up a hobby (painting, French cooking), attending a museum or cultural performance, or participating in civic procedures. PBS also pioneered telephone hotlines and exploited the postal system to implement viewer votes by mail on public affairs debates. Such experiments were steeped in the "art of effects", born of the "governmental" impulse to cultivate high-caliber citizens who participated in their cultural improvement and needed to be taught how to participate wisely in US democracy. Interactivity in the service of fundraising was developed simultaneously. Many PBS stations published membership magazines or newsletters with programming-related articles, reviews, and station news and publicity. Sent only to contributors, the magazines encouraged a virtual community around stations and PBS. Televised auctions and pledge drives cultivated a similar sense of belonging and harnessed the telephone as an instrument through which television viewers could immediately "take action" in support of the institution that looked out for their cultural welfare. In this sense, goals of cultivating citizens and fundraising were conjoined as a form of ethical consumerism. Both approaches to interactivity have provided rudimentary templates for the private television industry's subsequent development of the web and other new technologies as supplemental resources, participation venues, and marketing mechanisms. Likewise, PBS now takes cues from commercial channels as it pursues residual and emerging forms of interactivity through the web.

The initial approach to interactivity as a relay between PBS and practices of good citizenship can still be found on PBS.org and its associated websites. However, the "art of effects" has been tempered by a quest to attract consumers with many online options. Educational materials are now more apt to incorporate entertainment or take the form of interactive games, as exemplified by *The News Hour*'s interactive 2008 election map, which allowed users to "predict" election winners based on clickable information, providing opportunities for pleasure and agency that go beyond the mastery of a civic tutorial. The notion that citizens must be served and reformed to better fulfill national "duties and obligations" has also been tempered with an emphasis on providing do-it-yourself educational and lifestyle resources. Across PBS sites, citizens are addressed as active consumers of the information needed for successful living and self-care, from *History Detectives* (how to identify and value an artifact in your attic) to PBS Parents (how to get your child to eat vegetables). This parallels the advice programming increasingly featured on PBS, particularly during pledge season, when executives are especially concerned to connect with audiences. While this focus on self-service citizenship is common across media in an epoch of increased individualization and privatization, it also supports PBS's business goals. Not only is web activity needed to measure the value of PBS programs, it also provides additional revenue streams. Since the web does not come under FCC jurisdiction, PBS is able to sell advertising banners and paid sponsor links, opening up a whole new frontier for cross-platform marketing. Courting online users with interactive resources that feel fun and self-empowering, as opposed to a civic "duty" to be carried out under public television's tutelage, is one way of bolstering PBS's market value as a high "points of impact" brand.

Interactivity as fundraising has also thrived online, where it has been joined by intensified promotional and branding strategies. The web has become a vehicle for selling a proliferating array of PBS merchandise, including videos and tie-in books, mugs, and so on. Clickable opportunities to pledge money to public television also permeate virtually every layer of the interactive PBS experience. More efficient than local pledge drives (many auctions have migrated online for similar reasons), interactive fundraising enlists web users in what Mark Andrejevic calls the offloading of marketing in the flexible online economy. Individuals are encouraged not only to donate funds, but to take on the labor of making themselves into PBS viewers and members by filling out forms, providing information about themselves, and making a personal commitment to public television without being prodded by pledge hosts.[35] While the beckon to "give back" to PBS hinges on its residual role as a provider of cultural resources, online pledging is part of an array of self-branding activities carried in the spirit of its new strategic plan. For example, PBS's employment of user-generated video and social-networking websites ties interactive fundraising to participation and fandom. On PBS's YouTube channel, users are invited to watch teasers of current PBS programs, post and read comments, and follow links back to PBS in order to sign up as donors/members. Similarly, the official Facebook PBS fan site encourages visitors to fill out PBS surveys, participate in PBS polls, upload photos, and link their personal Facebook sites to PBS, in addition to pledging financial support. In both cases, the energy of interactive users is channeled into the health and vitality of PBS with minimal cost—the logic of efficiency.

YouTube and Facebook are also mobilized as interactive devices for building PBS fan communities—a tactic that permeates PBS's own websites as well. The *Masterpiece* site exemplifies a growing emphasis on pleasurable investment (rather than top-down uplift) by hosting a fan club, a discussion board, actor profiles, and even a recipe exchange. This suggests a partial rethinking of the negative judgment of television that gave rise to PBS. However, there are business reasons for encouraging fan dispositions. As Henry Jenkins points out, sponsors are increasingly looking to tap into viewers' affective investments with television programs—a possibility also suggested by PBS's point of impact metric.[36] Mobilizing people to give money to PBS similarly requires a strong investment in programming that involves feelings and pleasures, rather than learning alone. Pledge drives have historically sought to cultivate such affective loyalties—which is why "special" popular programs (such as concerts or old films) tend to be shown during pledge season. Online fandom turns every PBS program into an affective experience—and in so doing also makes viewers partners in PBS's new branding strategies. As Adam Arvidsson argues, any attempt to cultivate a feeling of community around a brand requires the active involvement of consumers. Brand managers who exploit old and new media to stimulate a sense of "managed community interaction" are fully aware of the extent to which consumers produce the brand's identity, he contends.[37] In lieu of PBS's entrepreneurial spirit, the web can be seen as a technology for cultivating market-friendly forms of citizenship as well as brand loyalists who produce "relevance"—and thus market value—for PBS through their online fan activity.

Following the lead of commercial broadcasters, PBS has begun to allow viewers to participate directly in some programming: *History Detectives*, for example, allows viewers to post questions for investigation on the program. However, this is limited in comparison to the participation encouraged by European public broadcasters. The lingering aura surrounding PBS does not allow it to pursue mass audience participation along the lines of *American Idol* or *Test the Nation*, a popular interactive quiz program shown on public broadcasting around the world, but passed over by PBS and shown on FOX as *Are You Smarter than a 5th Grader?*[38] PBS has also been reluctant to allow or encourage user-generated content on its websites. In only a few cases are viewers allowed to use the digitalized PBS infrastructure to circulate amateur journalism or cultural productions, along the lines of Current TV, a cable channel/website. The Corporation for Public Broadcasting, which oversees PBS, has, however, moved toward viewer involvement in a new branding campaign called My Source. In the campaign, real people appear in video testimonies explaining what they like best about public television. Posted to the web and shown on television, the spots look like YouTube videos. Emotional, affective, and reliant on the labor of the fans who appear in them, they are being used to recruit new PBS viewers and boost station memberships. Such are the possibilities—and limits—of the consumer orientation at work at PBS.

Self-Service Democracy: Public Service in the Ownership Society

While PBS was reinventing itself as a business and a brand, its public service mission was entrusted to the Digital Future Initiative, a private commission co-chaired by business consultant and former FCC chairman Reed Hunt and former Netscape CEO James Barksdale, president of the Barksdale Management Corporation. Comprised of prominent national business leaders, some of whom were also PBS board members, as well as some local PBS station managers, the commission—which was initiated by PBS president/CEO Pat Mitchell—was asked define the purpose and scope of public service television in the wake of ongoing technological developments. In 2004, it issued an influential report detailing the changing role of PBS (or its digital equivalent) as channels proliferate and the medium of television rapidly converges with the web, mobile phones, iPods, and other interactive delivery platforms. While the commission's recommendations are ultimately non-binding, the 2004 report exerts discursive influence as an unofficial policy document. Similar to the private Carnegie Commission's influential 1967 report on public television, it defines an agenda for the future and establishes the priorities and goals of public service television (and digital media) in a changing media environment.[39]

The relegation of cultural policy to the private sector has a long history in the United States: the Carnegie Commission on Public Television, which provided the blueprint for the US public television system, was also a hand-picked panel of business and cultural elites who performed their duties outside public jurisdiction. This isn't surprising for, as Thomas Streeter points out, US liberalism (and neoliberalism) has always located the "public interest" within corporate capitalism, not outside it.[40] However, whereas the Carnegie Commission was unable to reconcile the values of liberal democracy with commercial television of the 1960s and recommended an alternative non-commercial program service to provide "excellence and diversity," the Digital Future group did not make this claim. Presuming that the expanded television marketplace now provides excellence and diversity, its report detailed other justifications for public and private funding for the digital equivalent of PBS.

In the interventionist milieu of the 1960s, tax-based funding was authorized in the hopes that public television could be mobilized to raise human capacities across all spheres of life through an informal liberal arts and sciences curriculum. While bound to class hierarchies and the "art of effects," this mission also rooted PBS in the welfare stage of liberal capitalism, characterized by Nikolas Rose as a social contract in which the state offers "rights to social protection and social education in return for duties of social obligation and social responsibility."[41] As many observers have shown, the expanded social contract of the 1960s has been under reform for some time, as policymakers and critics across the political spectrum claim that too much reliance on the public sector has bred inefficient public sector bureaucracies as well as overly dependent citizens. Since the 1990s, particularly in the United States, there has been a wave of privatization measures and efforts to downscale welfare programs initially designed to compensate for unequal opportunities, as well as the State's retreat from public education—particularly in the liberal arts and sciences—paving the way for culture wars and the rise of the corporate university. There have also been both the "zeroing out" of need-based welfare programs and a downscaling of earlier efforts to equalize educational opportunities. As this has happened, self-help has exploded, and politicians and media alike have helped introduce an ethic of personal responsibility, with citizens being called upon to secure their own fates and mobilize their own social and cultural resources.[42]

Within this context, the Digital Future Initiative cautiously identified residual and new public service goals. Civic participation, always emphasized by PBS, was briefly cited as a continued priority. The 2004 report briefly discussed PBS's proposed "Public Square" project, an attempt to expand national public affairs to make public television a home for "social capitalists." Public Square was to feature "in-depth news, thought-provoking discussions, and challenging documentaries" as well as viewer interaction, and was billed as a "dynamic, vibrant national civic space in which citizens are informed about and engaged with the world around them." The report recommended that local stations develop similar civic projects using multicasting channels, funded by private sources on account of the potentially "controversial" nature of journalism and public affairs. Tellingly, neither the national project nor the local versions have materialized.

This failure speaks to the changing ideals of citizenship in the United States—particularly the growing emphasis on lifestyle and self-care discussed in the previous section—and to the inability to secure resources for communal forms of civic participation. When PBS was unable to secure adequate funding for Public Square, the project was downsized to the PBS Engage website, which—in sync with the interactive branding efforts discussed earlier—defines participation as the chance to read blogs written by PBS commentators and connect with favorite PBS shows (the possibility of "remixing" PBS programs is in the works). The site does not accept user-generated content or allow users to set an agenda for discussion, and in that sense does not stray too far beyond public television's traditional focus on disseminating and mediating expert commentary. Meanwhile, only a few PBS stations offer local digital civic channels, and those that do—such as KTCA's Minnesota Channel (Minneapolis-St. Paul)—tend to use them as platforms for programs supplied by local businesses and educational institutions.

Universal access is another issue where the forces of privatization tend to override earlier conceptions of public service. While PBS has skewed upward in terms of its core demographics, the idea that public television would provide cultural resources to all Americans was integral to its creation as a partly publicly funded institution. This commitment continues to this day. Across the nation, many PBS stations are running on-air and newspaper campaigns to educate viewers about the upcoming transition to digital television and coaching viewers on the federal vouchers and new equipment they will need to continue receiving free television programming. Looking to the new media future, the 2004 report advises public broadcasters to see themselves less as broadcasters than as gateways or "interfaces" to a vast archive of public service content to be accessed and downloaded on demand.[43] Operating as "trusted" national brands rather than as non-commercial distributors per se, such public communications agencies as PBS would partner with commercial ventures to make a wealth of indexed and customizable public service content accessible "anytime, anywhere."

While the Public Service Media Search Engine recommended by the report has not yet materialized, PBS has created dispersed mechanisms for downloading existing PBS programs on a fee basis, in partnership with commercial distributors such as Google Video and iTunes. PBS has also made some of its content available on digital streaming sites such as Joost and Hulu, which require users to watch commercials as a condition of access. Some of its online sites have implemented subscription-tier services, including PBS Kids, which, for a monthly fee, offers access to PBS shows as well as interactive games.[44] In many respects, this trend simply builds on PBS's own merchandising activities, where the private "ownership" of public television, created in part through public funding, has long been valorized. However, the prospect of paying for brand-name public service codified by the Digital Future report sets a precedent for how universal access is conceived in the new media era. Some PBS programs can currently be viewed free of charge online (presuming the user can pay for broadband access): for example, *Frontline*'s website allows online users to watch documentaries from the series, including *Bush's War*, which has received far more attention on the web than during its PBS broadcast. Contrary to the BBC, which developed a free archive of its programming in accordance with its digital public service mission, many PBS programs cannot be streamed without incurring the fees and costs (including the labor of watching commercials) associated with commercial distribution, which effectively requires individuals to pay for the cultural resources associated with common citizenship. Such are the constraints of public service in what George W. Bush calls the Ownership Society.

The Digital Future commission identified school services and digital education to children, their parents, and caregivers as additional public service goals. Building on the mission established by ETV in the 1950s, PBS has always prioritized educational children's programming. However, the equality of opportunity adopted during the Great Society era has waned, as corporations, parents, and caregivers are now called upon to take responsibility for early childcare education. PBS currently supplies curriculum resources directly to schools, and do-it-yourself learning materials to individuals via the PBS Kids and PBS Parents websites, as a partner in the federal "Ready to Learn" program. Corporate sponsors partner in these PBS ventures as well, in exchange for banner ads, sponsored links, underwriting credits, and prominent placement in curriculum materials. The 2004 report advises extending these instructional partnerships to combat the "literary and learning crisis" with digital technologies that "can reach people everywhere they are with educational content, customized delivery on devices that occupy more time than school." This is not defined as a public service imperative in the old sense of using television to provide education (as a component of welfare) for all. The delivery of education via interactive and personalized media is justified as necessary for training the "high quality workforce essential for national competitiveness in the global economy."[45] Toward that end, the report recommends that public television pursue "interactive educational media" partnerships with firms such as LeapFrog Enterprises, a commercial educational software company, and the US military, pioneer of simulation technologies used to train soldiers.

To the extent that public education in the United States has been increasingly privatized and outsourced to corporate sponsors, the Digital Future Initiative merely positions the next stage of public media as an integral component of these trends. The report does not invent the changing ethic of welfare as much as it assures public television's ability to adapt accordingly as the basis for public as well as private funding. The commission's promotion of do-it-yourself education via the web and other new technologies speaks to the new demands being placed on citizens and the heightened role of self-help as a technology of welfare.

Tellingly, the final goals identified by the Digital Future report emphasize the need to train citizens to manage their own fates and futures in insecure times. This emerging definition of public service moves us even further from the idealist 1960s vision of PBS as a diffuser of liberal arts and sciences through a common curriculum. Defining the future equivalent of PBS not as a cultural service at all, but as a conduit for narrowly conceived (and increasingly technical) self-help information, the report ultimately equates the future of public service with the individualized management of personal risk and security.

While PBS has always promised to facilitate cultural and lifestyle pedagogy, its digital equivalent is called on to provide on-demand access to a narrow range of practical information and usable skills. This includes "lifelong learning opportunities" for contemporary US workers, who, in the absence of job security or a social safety net, require "more skills and knowledge than ever before to qualify for and keep new jobs."[46] Likewise, the future equivalent of PBS is called upon to solve the mounting US public healthcare crisis by providing access to an array of personal resources for self-monitoring and self-care. The crisis is characterized not as unavailability of affordable healthcare but as the rise of self-induced "diseases of lifestyle," such as smoking and obesity, which force insurance companies to raise their premiums. In light of this assessment, public service is said to entail operating as a "trusted" source or brand of self-help resources and personalized health information, extensively archived and delivered using the latest technologies.[47] To accomplish this, PBS or its digital equivalent would partner with the Centers for Disease Control and leading university medical schools, along the lines of the Health of Kentucky Initiative—a broadcast series and multimedia toolkit produced by Kentucky Educational Television—or WXXI's Second Opinion project in Rochester, New York, a multi-platform initiative co-produced by a local hospital, to "let Americans take charge of their own healthcare by giving them the information they need to make informed decisions." What is striking about this interpretation of the future of public service is not only how narrow these goals seem compared with the cultural ideals of the past, but how synchronized they are with the post-welfare expectation that today's citizens solve their own problems using the dispersed resources available to them. When public service is envisioned as a technology for equipping individuals to overcome diminished material resources, it is not surprising that the cultural enlightenment goals of the 1960s—such as exposing citizens to the arts or mobilizing them to vote on televised debates—seem less significant or urgent.

The report concludes by identifying emergency preparedness as an emergent public service opportunity. The system's infrastructure is offered up to the threat of terrorist attacks and other crises, such as Hurricane Katrina, where a "failure of communication" is cited as the cause of devastation and disaster (the lack of public investment in New Orleans before or after the hurricane is not mentioned). In partnership with federal and state networks, PBS and its member stations are enlisted to foster emergency communication among first responders and make available preparedness education distributed not only through broadcast channels but through web archives, data casting, schools, DVDs, websites, and other old and new technologies. For providing this, the report suggests public television should receive subsidies from the Department of Homeland Security and other federal agencies.[48] As Mark Andrejevic and James Hay have argued, the Department of Homeland Security, a uniquely sanctioned federal domestic program in the age of downsized government, offloads much of the responsibility for "readiness" onto individuals.[49] To the extent that personal risk management is what counts as welfare in the United States today, redefining the oasis of the wasteland as an instrument of securitization is a telling indication of where digitalized public service television is heading.

While the Digital Future Initiative calls for public funding (except for public affairs) to facilitate the new stage of public television, it looks to the private sector for the bulk of the resources needed to reinvent public service for the new digital age. Continuing the tradition of public–private partnership, it calls on philanthropies and corporations to open their pocket books to provide the vast archive of public service materials needed to create tomorrow's useful and self-sufficient (healthy, skilled, and prepared) citizenry. However, investment in new branding and fundraising strategies continues to be important, as each individual is called upon to contribute financially to the digital future as well: "From the child who collects pennies to donate to her local station to the mega-donor who seeds the Digital Future Endowment," the subjects of public service are simultaneously conceived as demanding consumers and citizen "partners for building a brighter future for this country."[50]

Concluding Thoughts: The Logic and Limits of Reinvention

PBS's efforts to reinvent itself as a channel, a brand, and a resource of citizenship are contradictory. On the one hand, the entrepreneurial spirit has prompted a partial rethinking of some of US public television's cultural biases and top-down priorities. Yet, PBS has mobilized the "art of effects" for new institutional purposes that ultimately reify its long-standing devotion to upscale, professional lifestyle clusters. The attempt to redefine public service in the context of new technologies and public television's own accelerated marketing and branding raises questions about the future of citizenship for constituencies who don't count as consumers. The positioning of PBS as a national network, trusted brand, and, increasingly, archivist of customized self-help information is, moreover, not so unlike what the television industry now offers to a broader audience. Increasingly, commercial ventures from

The Biggest Loser (NBC) to *Extreme Makeover: Home Edition* (ABC) integrate national television, interactive marketing, and multi-platform resources for pursuing health and nutrition, home ownership and security, and volunteerism. Endorsed by politicians (including George W. Bush), these corporate helping ventures rely on public–private partnerships with government agencies and both professional and charitable organizations and provide a branded gateway to a privatized, self-service democracy.[51] Whether PBS or its digital equivalent will recognize broader possibilities for public service remains to be seen—but the prognosis isn't promising. Certainly, the future of PBS—the only ostensibly non-commercial television platform the United States has ever seen—deserves to be debated beyond the confines of the strategic plans and private reports detailed here.

Notes

1 A. J. Frutkin, "Reinventing PBS," *MediaWeek*, November 4, 2002, http://www.mediaweek.com/mw/research/article_display.jsp?vnu_content_id=1754091; Joseph Weber, "Public TV's Identity Crisis," *Business Week*, September 30, 2002, http://www.businessweek.com/magazine/content/02_39/b3801087.htm; Karen Everhart Bedford, "PBS Goals: Reverse Slides in Audience, Membership," *Current*, April 23, 2001, http://www.current.org/pbs/pbs0108.html; Slyvia M. Chan-Olmsted and Yungwook Kim, "The PBS Brand versus Cable Brands: Assessing the Brand Image of Public Television in a Multichannel Environment," *Journal of Broadcasting & Electronic Media*, 46, 2, 300–20.

2 *Digital Future Initiative: Challenges and Opportunities for Public Service Media in the Digital Age* (Alexandria, VA: PBS Foundation, 2004).

3 Georgina Born, "Digitalising Democracy," in *What Can Be Done? Making the Media and Politics Better*, special book issue of *Political Quarterly*, ed. J. Lloyd and J. Seaton (Oxford: Blackwell, 2006), 102.

4 See also Georgina Born, *Uncertain Vision: Birt, Dyke and the Reinvention of the BBC* (London: Secker & Warburg, 2004).

5 Laurie Ouellette, *Viewers Like You? How Public TV Failed the People* (New York: Columbia University Press, 2002).

6 Graham Murdock, "Public Broadcasting and Democratic Culture: Consumers, Citizens and Communards," in *A Companion to Television*, ed. Janet Wasco (Malden, MA: Blackwell, 2005), 186.

7 Pierre Bourdieu, *Distinction: A Social Critique of the Judgment of Taste* (Cambridge, MA: Harvard University Press, 1984). For a more sustained discussion of the politics of taste and culture in relation to PBS, see Ouellette, *Viewers Like You?*

8 The rationality of public sector reform in the United States is spelled out in David Osborne and Ted Gaebler's influential book *Reinventing Government: How the Entrepreneurial Spirit is Transforming the Public Sector* (New York: Plume, 1992), which was endorsed by Bill Clinton and has been widely cited as a blueprint for reform.

9 Frutkin, "Reinventing PBS"; Susan E. Linn and Alvin F. Poussaint, "The Trouble with Teletubbies: The Commercialization of PBS," *American Prospect*, May/June 1999, http://www.prospect.org/cs/articles?article=the_trouble_with_teletubbies. While Mitchell left PBS in 2006, the reinvention strategies she helped launch continue to reverberate.

10 John Consoli, "PBS Hones its Pitch with New Sponsorships," *MediaWeek*, September 25, 2006, http://www.mediaweek.com/mw/research/article_display.jsp?vnu_content_id=1003155605.

11 Mark J. Pescatore, "More Commercial Behavior from PBS," *Government Video*, October 1, 2006, 4.

12 For a political economic critique of the escalating commercialism of PBS in the late 1990s, see William Hoynes, "The PBS Brand and the Merchandising of Public Service," in *Public Broadcasting and the Public Interest*, ed. Michael P. McCauley, Eric E. Peterson, B. Lee Artz, and DeeDee Halleck (Armonk, NY: M. E. Sharpe, 2003), 41–51.

13 John Caldwell, "Convergence Television: Aggregating Form and Repurposing Content in the Culture of Conglomeration," in *Television After TV: Essays on a Medium in Transition*, ed. Lynn Spigel and Jan Olsson (Durham, NC: Duke University Press, 2004), 41–74.

14 Ibid.

15 Justin Lewis, *Art, Culture and Enterprise* (London: Routledge, 1990).

16 *Current* Briefing, "What Pubcasters are Doing to Increase Impact and Reduce Operating Costs," http://www.current.org/mo/mo3.html; Frutkin, "Reinventing PBS."

17 Karen Everhart, "I'm Staying to Do Whatever I Can: Mitchell Renews her Commitment," *Current*, October 20, 2003, http://www.current.org/pbs/pbs0319mitchell.shtml.

18 Frutkin, "Reinventing PBS."

19 Ibid.

20 The need to attract viewers who are also donors adds another dimension to the institutional imperatives and anxieties surrounding the amorphous and ultimately unknowable audience addressed by Ien Ang. See *Desperately Seeking the Audience* (London: Routledge, 1991).

21 Weber, "Public TV's Identity Crisis."

22 Mae Anderson, "Fallon Evolves PBS Makeover," *Adweek*, September 8, 2003, 34.

23 John Consolik, "PBS Hones its Pitch with New Sponsorships," *MediaWeek*, September 25, 2006.

24 Stephen Perry, "PBS Utilizes Multicasting Capabilities with Create," *Government Video*, August 1, 2006, 6.

25 Elizabeth Jensen, "Public Television Plans a Network for Latinos," *New York Times*, February 7, 2007, E1; Arian Campo-Flores, "Latino TV Gets Serious," *Newsweek*, March 19, 2007, 48.

26 "PBS–Comcast Kids Channel Exemplifies Public–Private Partnership, Says PBS," *Public Broadcasting Report*, April 15, 2005; "PBS Keeps Focus on Preschoolers," *Television Week*, April 7, 2008, 13.

27 Concerns about the escalating commercialism of PBS emerged in the 1990s: See James Ledbetter, *Made Possible by ... the Death of Public Broadcasting in the United States* (New York: Verso, 1997), and William Hoynes, *Public Television for Sale* (Boulder, CO: Westview Press, 1994). For a more recent critique of the marketization of public broadcasting as an assault on the public interest and citizenship, see McCauley *et al.*, *Public Broadcasting and the Public Interest*.

28 I discuss the contradictions of this mission at length in *Viewers Like You?* See also Laurie Ouellette, "TV Viewing as Good Citizenship? Political Rationality, Enlightened Democracy, and PBS," *Cultural Studies*, 13(1) (1999), 62–90.

29 Ang, *Desperately Seeking the Audience*, 103; Robert Blakely, *To Serve the Public Interest: Educational Broadcasting in the United States* (Syracuse, NY: Syracuse University Press, 1979), 28. Ang alluded to but did not elaborate on Foucault's work on governmentality, which has only recently been addressed by media scholars. For an introduction to the concept, see Michel Foucault, "Governmentality," in *The Foucault Effect: Studies in Governmentality*, ed. Graham Burchell, Colin Gordon, and Peter Miller (Chicago: University of Chicago Press), 87–104; Colin Gordon, "Governmental Rationality: An Introduction," ibid., 1–54; and Graham Burchell, "Liberal Government and Techniques of the Self," in *Foucault and Political Reason: Liberalism, Neo-Liberalism and Rationalities of Government*, ed. Andrew Barry, Thomas Osbourne, and Nikolas Rose (Chicago: University of Chicago Press, 1996), 19–36. The "art of effects"

can also be a component of commercial programming. While this issue is beyond the scope of this essay, it is taken up extensively in Laurie Ouellette and James Hay, *Better Living Through Reality TV: Television and Post-Welfare Citizenship* (Malden, MA: Blackwell, 2008).

30 Ang, *Desperately Seeking the Audience*, 28–9; 106.

31 Everhart Bedford, "PBS Goals."

32 Lynn Smith, "Knickers in a Twist: Masterpeice Theatre is Being—Egad—Rebranded to Appeal to a Younger Crowd," *Los Angeles Times*, January 11, 2008, E1.

33 Quoted in Frutkin, "Reinventing PBS." PBS makes similar claims in its sponsorship materials, available online at www.pbs.org.

34 Frutkin, "Reinventing PBS."

35 Mark Andrejevic, *iSpy: Surveillance and Power in the Interactive Era* (Lawrence: University of Kansas Press, 2008).

36 Henry Jenkins, "Buying into American Idol: How We are Being Sold in Reality TV," in *Convergence Culture: Where Old and New Media Collide* (New York: New York University Press, 2006), 59–92.

37 Adam Arvidsson, *Brands: Meaning and Value in Media Culture* (London: Routledge, 2006).

38 PBS's ties to niche marketing and refined cultural sensibilities complicate Gunn Sara Enli's inclusion of *History Detectives* as a broader global trend toward inviting viewer participation in public broadcasting, as evidenced by the wide adoption of *Test the Nation*. See "Redefining Public Service Broadcasting: Multi-Platform Participation," *Convergence: A Journal of Research into New Media Technologies*, 14, 1 (2009), 105–20.

39 The Carnegie Commission released its own report, *The Meeting of Two Cultures: Public Broadcasting on the Threshold of the Digital Age* (New York: Carnegie Corporation of New York, 2008) just before this book went to press.

40 Thomas Streeter, *Selling the Air: A Critique of the Policy of Commercial Broadcasting in the United States* (Chicago: University of Chicago Press, 1996).

41 Nikolas Rose, "Governing 'Advanced' Liberal Democracies," in *Foucault and Political Reason*, 49.

42 See Ouellette and Hay, *Better Living Through Reality TV*, for an extended discussion of this development.

43 Celia Lury uses the term "interface" to emphasize the extent to which brands operate similarly to new media objects on account of their interactive dimensions. The term aptly describes the vision of PBS as a branded gateway or entry point into customized and downloadable information rather than a traditional broadcaster. See Celia Lury, *Brands: The Logos of the Global Economy* (London: Routledge, 2004).

44 Alex Weprin, "Hulu to Add PBS Shows," *Broadcasting & Cable*, June 10, 2008; Mike Shields, "PBS Launches Joost Channel," *MediaWeek*, December 18, 2007; Alicia Zappier, "PBS Pilots New Program Initiative with Google Video," *Television Broadcast*, September 1, 2006, 8; Warren Buckleitner, "Children's Games and Shows by Subscription, from PBS," *New York Times*, March 13, 2008, 7.

45 *Digital Future Initiative*, 6.

46 Ibid., 7.

47 Ibid., 80.

48 Ibid., 89.

49 James Hay and Mark Andrejevic, "Introduction to Special Issue on Homeland Insecurities," *Cultural Studies*, 20, 4–5 (2006), 331–8.

50 *Digital Future Initiative*, 114.

51 See Ouellette and Hay, *Better Living Through Reality TV*, for an analysis of reality and lifestyle television as teachnologies of welfare.

Index

Also from Routledge

Production Studies: Culture Studies of Media Industries

Co-edited by Vicki Mayer, Miranda J. Banks, and John Thornton Caldwell

"Behind-the-scenes" stories of ranting directors, stingy producers, temperamental actors, and the like have fascinated us since the beginnings of film and television. Today, magazines, websites, television programs, and DVDs are devoted to telling tales of trade lore—from on-set antics to labor disputes. The *production* of media has become as storied and mythologized as the *content* of the films and TV shows themselves.

Production Studies is the first volume to bring together a star-studded cast of interdisciplinary media scholars to examine the unique cultural practices of media production. The all-new essays collected here combine ethnographic, sociological, critical, material, and political-economic methods to explore a wide range of topics, from contemporary industrial trends such as new media and niche markets to gender and workplace hierarchies. Together, the contributors seek to understand how the entire span of "media producers"—ranging from high-profile producers and directors to anonymous production assistants and costume designers—work through professional organizations and informal networks to form communities of shared practices, languages, and cultural understandings of the world.

This landmark collection connects the cultural activities of media producers to our broader understanding of media practices and texts, establishing an innovative and agenda-setting approach to media industry scholarship for the twenty-first century.

Contributors: Miranda J. Banks, John T. Caldwell, Christine Cornea, Laura Grindstaff, Felicia D. Henderson, Erin Hill, Jane Landman, Elana Levine, Amanda D. Lotz, Paul Malcolm, Denise Mann, Vicki Mayer, Candace Moore, Oli Mould, Sherry B. Ortner, Matt Stahl, John L. Sullivan, Serra Tinic, Stephen Zafirau

Vicki Mayer is Associate Professor of Communication at Tulane University. She is author of *Producing Dreams, Consuming Youth: Mexican Americans and Mass Media* as well as *Below the Line: Producers and Production Studies in the New Television Economy*.

Miranda Banks is Assistant Professor of Visual and Media Arts at Emerson College.

John Thornton Caldwell is Professor of Film, Television, and Digital Media at UCLA. He has authored and edited several books, including *Televisuality: Style, Crisis and Authority in American Television, New Media: Digitextual Theories and Practices*, and *Production Culture: Industrial Reflexivity and Critical Practice in Film and Television.*

ISBN10: 0–415–99796–8 (pbk)
ISBN10: 0–415–99795–X (hbk)

ISBN13: 978–0–415–99796–6 (pbk)
ISBN13: 978–0–415–99795–9 (hbk)